# REINFORCED PLASTICS

## Properties & Applications

# REINFORCED PLASTICS

## Properties & Applications

**Raymond B. Seymour**
Distinguished Professor of Polymer Science
University of Southern Mississippi

**The Materials
Information Society**

Library of Congress Catalog Card No.: 91-71393
ISBN: 0-87170-414-5
SAN: 204-7586

Editorial and production coordination by
Kathleen Mills Editorial & Production Services

PRINTED IN THE UNITED STATES OF AMERICA

*This book is dedicated to the many polymer scientists with whom I met and exchanged ideas at the Gibson Island Conferences in the late 1930s, the later Gordon Conferences, the annual meetings of the Society of Plastics Engineers, and the polymer-oriented divisions of the American Chemical Society.*
*I especially wish to acknowledge the contributions of Drs. William Bailey, Daniel Fox, Paul Flory, Carl Marvel, and Frank Mayo, who made numerous contributions to polymer composite science during their productive lives.*

# Preface

Most animals and plants are naturally occurring composites, and the relationships among the various components of these composites have been investigated for centuries. In contrast, most synthetic polymers, such as fibers, uncompounded elastomers, and unfilled plastics, are not composites. Their utility depends, to a large extent, on entanglement of the polymer chains and intermolecular attractions among these chains.

It is interesting to note that celluloid, the first man-made plastic, was a composite consisting of intractable cellulose nitrate and camphor, which served as a flexiblizer or plasticizer. Likewise, Bakelite, the first truly synthetic plastic, was a composite consisting of a phenolic polymer reinforced by a wood flour filler.

Since early developments in plastic technology were empirical and general-purpose thermoplastics such as polystyrene, polymethyl methacrylate, polyethylene, and polypropylene (commercialized in the 1930s and 1950s) were less dependent than celluloid and Bakelite on additives, composite science was neglected until a few decades ago.

Polyvinyl chloride, also a pioneer general-purpose plastic, had limited use until it was flexibilized by the addition of phthalic acid ester plasticizers. The plasticized PVC (Koroseal) served well as a flexible plastic, but the utility of PVC was increased dramatically when heat stabilizers were added in the 1940s. Likewise, because it was brittle, the use of polystyrene was somewhat restricted until it was toughened by the addition of an elastomer, such as styrene-butadiene rubber.

Despite their wide use, most general-purpose plastics could not function as components of aircrafts, boats, automobiles, or recreational equipment. Fortunately, advances in composite science have resulted in the production of plastic composites, which extend the usefulness of plastic far beyond that of the general-purpose thermoplastics. Fortunately, we are now living in the Age of Composites, in which the performance of both thermosets and thermoplastics is enhanced by compounding with properly selected additives.

The present state of the art of compounding of plastics is described throughout this book. Since it was written primarily for designers and engineers, the emphasis is on useful combinations of plastics and appropriate additives. Some essential theories are discussed and references to more complex theories are provided.

A large number of engineers are now employed in the composites field; this number should double in the next few decades as the annual use of composites increases from a few million tons to more than 20 million tons. It is hoped that this book will help continue the growth of plastic composites. The Composites Age will be a golden age for knowledgeable engineers.

Raymond B. Seymour
Hattiesburg, MS

# Contents

# 11. REINFORCED THERMOSETS, 189

# 12. TESTS FOR PLASTIC COMPOSITES, 203

# APPENDIX 1
# U.S./SI UNITS: DEFINITIONS & CONVERSIONS, 213

# CHAPTER 1
# The Genesis of
# Plastic Composites

## INTRODUCTION

Engineers, scientists, and technologists educated prior to World War II learned about early developments in their field by enrolling in a course on the history of science or technology. Unfortunately, most of these courses have been abandoned and replaced by more modern courses such as computer science. Nevertheless, as Goethe stated, "The history of science is science itself." It should not be ignored or overlooked.

The evolution of a new science is not a "Big Bang" event. Instead, developments generally occur at irregular intervals over a period of years, and continue into the future. A brief history of plastic composites will be outlined here to show the steps and missteps in their development. The compounding of materials has gradually evolved from a "black art" to at least a pseudoscience.

## HISTORY OF POLYMERS
## PLUS FUNCTIONAL ADDITIVES

Some functional additives, such as curing agents and stabilizers, occur naturally in hevea rubber and thus may be the pioneer additives for polymers. Proteins found in rubber latex contain sulfur, which acts as a cross-linking or curing agent when heated, and quebrachetol, which serves as a stabilizer. The concentrations of these functional additives plus that of the phenolic stabilizers formed in the smoke of burning wood, used for coagulation, were acceptable for making tlachtli balls (the predecessors of modern basketballs). However, these concentrations are insufficient for other applications, such as waterproofing of textiles and modern radial tires.

White lead ($2PbCO_3 \cdot Pb(OH)_2$, the first man-made pigment, served as a curing agent (drier) when it was used as a pigment in linseed oil coatings many centuries ago. The dream of the alchemist who developed the Dutch process, which was based on the corrosion of lead by vinegar and carbon dioxide, would be a nightmare for a modern chemical engineer. The utility of this metal salt as a drier or siccative was improved by the Egyptians, who made lime soap over 34 centuries ago. Heavy metal salts of organic acids continue to be used as driers, but their most important use as additives is as heat stabilizers in polyvinyl chloride.

Another ancient additive was tannin, used to tan hides. Nowadays, tannic acid has been displaced, to a large extent, by other crosslinking agents, such as chromium sulfate, and selected polymers are used in place of leather in shoes and many other applications.

Hayward's solarization process developed in 1838 for curing rubber with sulfur was improved by Goodyear, who substituted thermal energy for solar energy in 1939. This slow vulcanization (crosslinking) process was accelerated in the early 1900s when Oenslager added solid aniline derivatives, such as thiocarbanilide, to the mixture of sulfur and rubber. The principal accelerators used today are derivatives of 2-mercaptobenzo-thiazole (Captax). Ostromislensky cured natural rubber by the addition of benzoyl peroxide.

Prior to 2000 B.C., the Egyptians knew that crushed fruit, when stored in a warm place, produced an intoxicating drink, and the making of beer and wine was practiced before 1500 B.C. The carbon dioxide by-product of yeast fermentation (leaven) was used in the making of leavened bread. Leavening was also accomplished several centuries ago by the use of soda ash ($Na_2CO_3$), which was extracted from plant ashes.

The Solvay process, developed in 1865, produced sodium bicarbonate ($NaHCO_3$), which could be mixed with solid acid salts to produce baking powder. Carbon dioxide from both fermentation and baking powder has been used to produce cellular polymeric products. Schridowitz patented this foam-making process in 1914. Other physical blowing agents (PBA), such as Freon, and chemical blowing agents (CBA), such as azobiscar-bonamide, are also used as propellants for polymeric foam production. However, the use of Freon and other chlorine-containing propellants is being discontinued because of their adverse effect on the ozone layer in the outer atmosphere. Fortunately, the cellular plastics industry in developed nations is meeting the requirements of the Montreal Protocol, and few chlorine-containing PBA's will be used in the 21st century.

Antioxidants, which are present in rubber latex and in many plants, provided some insight into the role of stabilizers. This was demonstrated by Hofmann and Spiller in 1861 and 1865. While Murphy patented the use of phenolic compounds as stabilizers in 1870, little progress occurred in this phase of polymer technology until 1922, when Moreau and Dufraisse showed that amine and phenol derivatives, which were used as accelerators, also inhibited the polymerization of styrene. Aniline derivatives were patented for use as antioxidants in the early 1920s by Caldwell, Winkelman, and Gray. The use of classic antioxidants such as 2,6-ditert. butylphenol (BHT) continues, but these antioxidants have been replaced, to some extent, by higher molecular weight hindered phenols and synergistic mixtures of additives.

Phenyl salicylate (salol), which was synthesized in 1886, was used first as an enteric coating for medicinals which must pass through the stomach unchanged. After it was discovered that salol absorbed ultraviolet light, it was used as a sunscreen in suntanning lotions and later as an ultraviolet stabilizer in polymers. Since the active peroxide decomposer is 2-hydroxybenzophenone (2-HBP) formed via a photo-Fries rearrangement, many derivatives of 2-HBP have been used as ultraviolet stabilizers.

Amines, lead tetraethyl, and lead and antimony salts, which were the first heat stabilizers for PVC, were patented by I.G. Farbenindustrie, DuPont, and Doolittle of Union Carbide in 1928, 1930, and 1934, respectively. While the use of lead salts continues in some nonfood plastics, these stabilizers have been replaced, to a large extent, by organotin and barium-cadmium compounds.

Excavations near Nice, France, indicate that the hominids used fire 500,000 years ago. The ancients not only learned how to utilize the energy from sustained chemical combustion reactions, but also discovered that organic materials, such as wood, were combustible and that ceramics and metallic materials were noncombustible. They also learned that flames could be quenched by water, and fire engines that could squirt a stream of water on fires were introduced in Alexandria in 100 B.C. Parmelee automatic sprinkler systems were installed in buildings in the U.S. in 1878.

It was generally recognized that celluloid, cotton, rayon, polystyrene, and polyolefins were combustible. Nondurable flame retardants such as borax ($Na_2B_4O_7 \cdot 10H_2O$), semidurable retardants such as precipitated metallic oxides, like stannic oxide ($SnO_2 \cdot H_2O$), and so-called durable retardants consisting of antimony and titanium compounds (Erifon) were used to improve the flame resistance of cellulosic textiles.

It was mistakenly assumed that PVC, aminoplastics, nylon, and polyurethanes were flame resistant, but it is now known that these plastics and even polytetrafluorethylene are combustible at elevated temperatures. It was discovered that combinations of PVC and antimony oxide ($Sb_2O_3$) were more flame resistant than PVC, and this led to the use of synergistic combinations of chloroorganic compounds and antimony oxide as flame-retardant additives for other polymers.

Bromoorganic compounds were found to be much more effective as flame retardants than chlorocompounds, and several bromo compounds, including *tris*(2,3-dibromopropyl phosphate) (tris) were developed by Walter in 1951. However, the use of tris, which contained 1,2-dibromo-3-chloropropane (a mutagen), was banned in 1977. Dow introduced a flame retardant by the bromination of 2,2-*bis*(2-hydroxyphenyl) propane in 1962. This is now the major brominated flame retardant used by the U.S. plastics industry.

One of the most widely used flame retardants, alumina trihydrate (ATH), produces steam when heated, and acts both as a flame retardant and a smoke suppressant. Phosphorus-containing compounds are also widely used as flame retardants.

Water, which is present as a flexibilizing agent in bipolymers, was also used by the ancients for softening wood, leather, and straw. Nineteenth century inventors attempted to flexibilize cellulose nitrate by the addition of gum copal, hevea rubber, castor oil, linseed oil, and camphor. However, the first commercial flexible cellulose nitrate (Celluloid) was not produced until 1869, when Hyatt selected camphor as the most effective plasticizer.

The plasticization of PVC with dibutyl phthalate by Semon and co-workers in the late 1920s made possible the use of intractable PVC. His success led to the development of processing aids for PVC by Jennings in the early 1950s. Acrylates were used as processing aids by Hopkins in 1960. Copolymers of styrene and methyl methacrylate are also widely used as processing aids.

The need for external lubricants and mold release agents was readily met by the use of salts of heavy metals and organic acids. The addition of silicones as internal mold lubricants followed similar logic and availability. It also seemed logical to fill the surface voids in reinforced polyesters by the addition of finely divided immiscible polymers such as polyolefins. The use of solubility parameter concepts has catalyzed the use of other immiscible polymers, such as polyurethanes and polycaprolactam, as low-profile agents (LPA).

The first use of colorants predates that of many other additives. The first paintings, such as those in the Altimara caves (15,000 B.C.), were based primarily on colored earths and clays. The concentrations and quality of the nonpigment components increased as the painting art improved. Sap from *Rhus urishiol* was used as the resinous component 3000 years ago in China, and egg albumin and other naturally occurring binders were used by contemporary artists in the Western world. Many new synthetic colorants were synthesized in the 19th century after Perkin produced mauve in 1856. From a volume viewpoint, the production of white titanium dioxide by Rossi in 1912 was a major advance in pigment technology. Titanium dioxide is the most widely used pigment today.

The use of other additives, such as antimicrobials, antiblocking agents, antifoaming agents, antifogging agents, and antistats, was usually the result of the incorporation of a readily available material into the polymer or a semiscientific development, based on a knowledge of the problem to be solved. Thus, oleamide, silica, fatty acid esters, quaternary amine compounds, and biocides such as copper-8-hydroxyquinolate were introduced as antiblocking, antifoams, antifogging, antistatic, and antimicrobial additives. The fragrances used by the soap industry were also used to solve odor problems or to create new odors in the plastics industry.

## HISTORY OF PLASTICS
## PLUS FILLERS AND REINFORCEMENTS

Some flame retardants, such as ATH, and pigments are used in relatively large amounts in plastics and could be classified as functional fillers. However, most fillers originally were incorporated in polymers to reduce surface tackiness, increase hardness, and extend the more expensive resinous substances, such as bitumens. Filled asphalt was used by the Sumerians for the production of mastic flooring in 3000 B.C.

In 1851, Nelson Goodyear used zinc oxide as a filler in ebonite and Hyatt added fillers such as ivory dust to his Celluloid. Baekeland used wood flour to extend Bakelite, and John used alpha cellulose as a filler for urea plastics in 1920. Asbestos was also used as a filler by Baekeland, but the use of this filler is decreasing because of toxicity problems.

Cloisonné, which consisted of reinforcement of fine metallic ribbons or wires in resinous binders, was used for jewelry by the Egyptians. Powdered lead and iron alloys were used two centuries ago in printing inks, and aluminum-filled plastics have been electroplated since the late 1940s. Magnetic tapes and discs were patented by Poulsen in 1900; these composites are now essential for modern communication.

Attempts were made to reduce the antistatic properties of nonpolar polymers by the addition of carbon black in the 1950s, and aluminum fibers were added to reinforced polyesters to increase their electrical conductivity in 1977. The use of these and other conductive fillers, such as aluminum flakes, became mandatory in the 1970s for shielding from electromagnetic interference (EMI) and radio frequency interference (RFI) in business machines and computer housings.

The production of glass fibers was patented by Slayter and Thomas in the 1930s. These fibers were used to reinforce unsaturated polyester resins, which were patented by Ellis and Foster in the same decade. Fiberglass-reinforced polyester radomes were used in airplanes in 1938. Fiberglass, which is used at an annual rate of 400,000 tons in the U.S., is the major reinforcement for polymers.

Carbon fibers were produced by the pyrolysis of cellulose filaments by Swan and Edison in the 1890s. Today, most of the 2000 tons of carbon or graphite fibers used annually in the U.S. is made by the pyrolysis of stretched acrylic fibers or pitch. The Lear jet and the Space Shuttle, which are based on graphite-reinforced epoxy resins, are popular examples of the state of the art of these composites.

Other fibrous reinforcements, such as aluminum silicate (Fiberfrax), silica, gypsum (Franklin fiber), potassium titanate (Tismo, Fybrex), silicon carbide (Silar), polyester (PET), nylon, and aramids (Kevlar), were introduced in the 1950s and '60s by General Electric, J.M. Huber, Jim Carter Company, DuPont and Otsuka, and DuPont. These additives are discussed in subsequent chapters in this book.

## HISTORY OF
## MODERN ADVANCED PLASTIC COMPOSITES

Modern plastic composites were unknown prior to World War II. In 1989, 14,000 tons of advanced plastic composites at a cost of $3.9 billion were used worldwide. About 60% of these advanced composites were used by the aerospace industry.

It is anticipated that over 30,000 tons of these advanced composites valued at over $8 billion will be consumed in 1999. In addition to these reinforced thermosets, 750 tons of reinforced thermoplastics (RTP) valued at $17 million were used in the U.S. in 1989. It is anticipated that the use of RTP's will increase to over 200,000 tons valued at $200 million in 1999.

In addition to the consumption of advanced plastic composites discussed above, the consumption of thermoset polymers in composites (in thousands of tons) in the U.S. in 1989 was as follows: epoxy, 39; phenolic, 850; and unsaturated polyester, 782. The pattern of consumption of RTP's, in thousand tons, was as follows: nylon, 80; polycarbonate, 30; PET (polyester), 88; polypropylene, 84; and polystyrene, 24.

## NATIONAL PLASTICS CENTER AND MUSEUM

Unlike many other materials, which require considerable time, money, and effort by archeologists to assemble information on their ancient history, most polymer history is 20th century history. A record of these advances is on exhibit at the National Plastics Center and Museum at Leominster, MA. Located in a former school building, the Center includes an exhibit area and educational videos such as *Passing Through*, which will be viewed by about 20 million high school science students annually. Obviously, since the accent is on construction and transportation, considerable information on plastic composites is included in this video.

## REFERENCES

· R.D. Deanin and N.R. Schott, Ed., *Fillers and Reinforcements for Plastics*, Advances in Chemistry Series No. 134, American Chemical Society, Washington, DC, 1974
· R. Gachter and H. Muller, *Plastics Additives Handbook*, Hanser Publishers, New York, 1983
· H.S. Katz and J.V. Milewski, Ed., *Handbook of Fillers and Reinforcements*, Van Nostrand Reinhold, New York, 1978
· R.J. Martino, Ed., *Modern Plastics Encyclopedia*, McGraw-Hill, New York, 1990

- N.G. McCrum, C.P. Buckley, and C.B. Bucknall, *Principles of Polymer Engineering*, Oxford University Press, New York, 1988
- T.J. Reinhart, Ed., *Engineered Materials Handbook*, Vol. 1, *Composites*, ASM International, Materials Park, OH, 1987
- P.D. Ritchie, Ed., *Plasticizers, Stabilizers and Fillers*, The Plastics Institute, London, 1972
- R.B. Seymour, Ed., *Additives for Plastics*, Vol. 1 and 2, Academic Press, New York, 1978
- R.B. Seymour, in *Wiley Encyclopedia of Polymer Science and Engineering*, Wiley-Interscience, New York, 1985
- R.B. Seymour, in *Wiley Encyclopedia of Packaging Technology*, John Wiley & Sons, New York, 1986
- R.B. Seymour and R.D. Deanin, Ed., *History of Polymer Composites*, VNU Science Press, Utrecht, The Netherlands, 1987
- R.B. Seymour, *Plastics for Engineering Applications*, ASM International, Materials Park, OH, 1987
- R.B. Seymour, *Engineering Polymer Source Book*, McGraw-Hill, New York, 1990
- A.M. Thayer, Advanced Polymer Composites, *Chem. Eng. News*, Vol. 68 (No. 30), 1990, p. 37

# CHAPTER 2
# Property-Enhancing Additives

## INTRODUCTION

Because a polymer composite material is often defined as a combination of two or more materials in which the polymer (resin) and the nonresinous constituent (reinforcing element or filler) retain their identities, there is too little emphasis on plastic composites in which the nonresinous additive is a filler or reinforcing agent. Yet, functional additives are essential for plastic performance and are a very important segment of the modern plastics industry. These functional additives are discussed in alphabetical order in this chapter.

Although polymer technicians have been reticent to discuss the role of additives in the past, practically every commercial polymer, with the exception of fibers, is dependent on the presence of properly selected additives. In the absence of camphor, Hyatt's Celluloid would have been limited to its use as an explosive, and without sulfur Goodyear's India rubber would have little utility.

Fortunately, von Siemens developed an extruder in 1847 that could be used to mix gutta percha and additives and extrude the "compound" as an insulating coating for wire. This extruder was improved in 1885 by John Royle and is still in use today in the rubber and plastics industries. The compounding of polymers—i.e., the combining of polymer with appropriate additives—is now one of these industries most important functions.

The American plastics industry compounded 5.8 million tons of composite plastics in 1990; it is anticipated that this volume will increase to 8.7 million tons in 1994. The growth-rate pacesetters will be conductive, plasticized, and plastic blends. As shown by the data in Table 2.1, tolling/custom and color compounding will account for more than 90% of the U.S. composite plastics business.

9

Table 2.1  Volume of Composite Plastics (tons)

| Sector | Consumption, tons 1990 | 1994 |
|---|---|---|
| Toll/custom | 245,000 | 327,000 |
| Color | 277,000 | 445,000 |
| Reinforced/filled | 22,700 | 31,800 |
| Alloys/blends | 1,800 | 4,500 |
| TPE/blends | 3,600 | 4,100 |
| PVC/plasticized | 44,000 | 55,000 |
| Conductive plastics | 3,600 | 4,100 |
| **TOTAL** | 576,000 | 873,000 |

## ACCELERATORS

Hayward's "solarization" process and Goodyear's vulcanization process, in which the tackiness of hevea rubber was reduced by a reaction with sulfur, were very slow. Fortunately, Oenslager found that the vulcanization process could be shortened from 120 min at 140 °C to a few minutes at 93 °C by the addition of aniline derivatives, such as thiocarbanilide (($C_6H_5NH$)$_2$ C=S). Since this catalyst, which is called an accelerator in the rubber industry, reduced the "tack" which was essential in calendared fabrics, a small amount of toxic aniline ($C_6H_5NH_2$) was also used with the thiocarbanilide, zinc oxide, and sulfur additives.

Thiocarbanilide was replaced, at least in part, by diphenylguanidine (DPG) (($C_6H_5NH$)$_2$C=NH), and most accelerators were replaced by 2-mercaptobenzothiazole (MBT) after its discovery in the early 1920s. MBT ($C_6H_4NSCSH$), which was used under the code name of Captax, is produced by the reaction of carbon disulfide ($CS_2$), sulfur, and aniline ($C_6H_5NH_2$).

Soft rubber, such as that used for tire tread stock, is produced by heating rubber with about 3% sulfur and about 0.3% accelerator. Larger amounts of sulfur (25%) produce hard rubber (ebonite).

Other derivatives of MBT, such as benzothiazyl disulfide (Altax) and the zinc salt of MBT (Zetax) are less active than MBT and have less tendency to cause scorching of the mixture of rubber and additives (compounds).

N-cyclohexyl-2-benzothiazolesulfenamide (Durax, Santocure) is produced by the sodium hypochlorite oxidation of the cyclohexyl ammonium salt of MBT. Durax and N-oxydiethylene benzothiazole-2-sulfenamide (Amax) are delayed action accelerators, which are preferred for use in the curing of styrene-butadiene elastomers (SBR). About 35,000 tons of thiazole accelerators are used annually by the American rubber industry.

Dithiocarbamates ($R_2NCS_2^-,NH_2R_2^+$), which are produced by the reaction of carbon disulfide and secondary aliphatic amines, are extremely active vulcanization catalysts called ultraaccelerators. The commercially available bismuth (Bismate), cadmium (Cadmate), copper (Cumate), lead (Ledate), selenium (Selenac), dimethyl ammonium (Zimate), and piperidinium (Pip Pip) salts are used to a large extent as latex accelerators. They will cure rubber in the absence of sulfur at room temperature.

The dithiocarbamates are readily oxidized to thiuram disulfides ($(R_2NCS_2)_2$), which are also used as ultraaccelerators. Tetramethyl and tetraethylthiuram disulfide (Tuads), dipentamethylenethiraum hexasulfide (Sulfads), and tetramethylthiuram monosulfide (Unads), are commercially available. The latter is an ultraaccelerator but, unlike the disulfides, is not a crosslinking agent by itself. About 4,000 tons of ultraaccelerators are used annually by the American rubber industry.

## ACCELERATOR ACTIVATORS

Zinc oxide (ZnO), which was used as a filler by Charles Goodyear and other rubber compounders, is also an activator for rubber accelerators. It is used with magnesia (MgO) as a curing agent for neoprene. It is also standard practice to add fatty acids, such as stearic acid ($H(CH_2)_{17}COOH$), which form salts by reacting with the zinc oxide. Litharge (PbO), magnesia (MgO), calcium carbonate ($CaCO_3$), furnace blacks (C), and amines ($RNH_2$), are also used as activators or accelerators in rubber compounding formulations.

## ANTIBLOCKING AGENTS

Because most plastic films are nonconductors of electricity, they tend to stick together. This nuisance may be overcome by the incorporation of fatty acid amides, such as erucamide ($H(CH_2)_{20}CONH_2$) and oleamide

($H(CH_2)_8CH:CH(CH_2)_7CONH_2$), fatty acid esters, waxes, metallic salts of fatty acids, fine silica, and even polymers, such as polyvinyl alcohol (PVA) ($-CH_2CHOH-)_n$, silicones, starch, and fluoropolymers ($-CF_2CF_2-)_n$. Some of these agents are incompatible additives designed to migrate to the surface, where they reduce the coefficient of friction and tack. The effective use of these additives may be evaluated by measuring their coefficients of friction. Plastic sheets containing additives should have a low value (0.2 to 0.4) for the ratio of the force required to move one surface past the other to the total force pressing the two surfaces together.

While antiblocking or flatting agents are widely used to prevent polymer films from adhering to each other, some of these additives interfere with the action of processing aids such as fluorocarbon additives and may cause melt fracture in extruded films unless specific polymer processing additives are present. The effectiveness of vinylidene fluoride-hexafluoropropylene copolymer as a flexibilizer in linear low-density polyethylene (LLDPE) is reduced by the presence of other additives as well as antiblocking agents.

The dusting problem with finely divided antiblocking agents may be solved by adding these additives to the molten plastic. The use of about 1% of these additives is effective in preventing the sticking of rolls of films to each other.

## ANTIFOAMING AGENTS

Among the many factors that cause foaming of plastics or polymeric systems are high bulk viscosity (which prevents bubble rupture), high surface viscosity, nonequilibrium surface tension diffusion (Marangoni effect), repulsion of electric double layers, entropy repulsion, and reduced gas diffusion between bubbles (which destabilizes the film).

Silicone-polyether antifoaming agents, mixtures of nonionic surfactants, and fatty acid esters are used during the dyeing of polyester-amides, and acetylene glycols are used in polyvinyl chloride (PVC) plastisols, adhesives, and coatings, including water-based coatings. Selected antifoaming agents, such as polysiloxanes (silicones), are used to control the rate of foam formation in cellular plastics. Antifoaming agents are also used in polymeric oil drilling muds.

Flatting agents, such as silicates or silicones, also prevent films from sticking to each other. The term "denesting agents" has also been used to

| Table 2.2  Annual U.S. Use of Antifoaming Agents | |
|---|---|
| **Additive** | **Consumption, tons** |
| Metallic stearates | 15,000 |
| Fatty acid amides | 8,600 |
| Paraffin wax | 7,300 |
| Fatty acid esters | 6,000 |
| Polyethylene wax | 2,600 |
| **TOTAL** | **39,500** |

describe these functional additives. The annual U.S. consumption of these compounds, which also have other functions, is shown in Table 2.2.

## ANTIFOGGING AGENTS

The view of packaged products may be obscured by the condensation of water as droplets on polymeric films. The addition of fatty acid esters may prevent fogging by forming a continuous film of water or by imparting hydrophobicity to the film surface.

## ANTIOXIDANTS

F. Hofman and K. Gottlob, who used piperidine ($C_5H_{10}NH$) instead of thiocarbanilide as a rubber accelerator in 1912, discovered that this heterocyclic amine prolonged the life of rubber tires. It has been found that any organic compound will deteriorate when exposed to oxygen in the atmosphere, and that the rate of this deterioration (autoxidation) is accelerated in the presence of impurities, such as copper ions or hydroperoxides.

In 1946, Bolland and Gee explained the autoxidation of organic compounds (RH) by a mechanism similar to that used to explain addition chain polymerization. This mechanism has also been used to explain the autoxidation of polymers.

A free radical (R•) is produced in the initiation step:

$$RH \rightarrow R\bullet + H\bullet \qquad \text{(Initiation)}$$

A new free radical (ROO•) is produced in the propagation step, and this free radical abstracts a hydrogen atom from the polymer and produces another free radical which can then react with oxygen in a chain reaction:

$$R\bullet + O_2 \rightarrow ROO\bullet \qquad \text{(Propagation)}$$
$$ROO\bullet + RH \rightarrow ROOH + R\bullet$$

In an autocatalytic reaction in which carbonyl and hydroperoxide compounds are cleaved at moderate temperatures or in sunlight, new free radicals are produced. These free radicals may cause chain breaking or crosslinking of polymers:

$$ROOH \rightarrow RO\bullet + \bullet OH$$
$$2ROOH \rightarrow RO\bullet + ROO\bullet + H_2O \qquad \text{(Autocatalysis)}$$
$$2ROOR \rightarrow 2RO\bullet$$

As shown by the following equations, the decomposition of the hydroperoxides (ROOH) is accelerated by metal ions, such as copper, iron, and manganese, which undergo one electron transfer:

$$ROOH + Mn^+ \rightarrow ROO\bullet + Mn^{++} + H^+ \qquad \text{(Metal catalyzed)}$$
$$ROOH + Mn^{++} \rightarrow RO\bullet + Mn^+ + OH^- \qquad \text{(Oxidation)}$$
$$2ROOH \rightarrow ROO\bullet + RO\bullet + H_2O$$

As shown by the following equations, terminations will produce inactive and crosslinked polymeric products:

$$2ROO\bullet \rightarrow \text{inactive products} + O_2$$
$$R\bullet + ROO\bullet \rightarrow ROOR \qquad \text{(Termination)}$$
$$2R\bullet \rightarrow R\text{–}R$$

Several investigators in the early 1920s synthesized antioxidants for rubber by the reaction of aromatic amines ($ArNH_2$) and aliphatic aldehydes (RCHO). While the rubber companies showed no interest in producing these additives, the U.S. Army funded their production by U.S. Rubber Company so that the deterioration of gas masks could be retarded during storage.

One of the first antioxidants produced (phenyl beta-naphthylamine, or Agerite, $C_{10}H_7NHC_6H_5$) continues to be the major antioxidant used by the rubber industry. The American rubber industry consumes more than 85,000 tons of antioxidants annually; amine compounds account for 60% of this volume.

As shown by the following equation for hindered phenols, both the secondary aromatic amines and hindered phenols react rapidly with peroxy radicals in a chain transfer reaction, in which a less active free radical is produced:

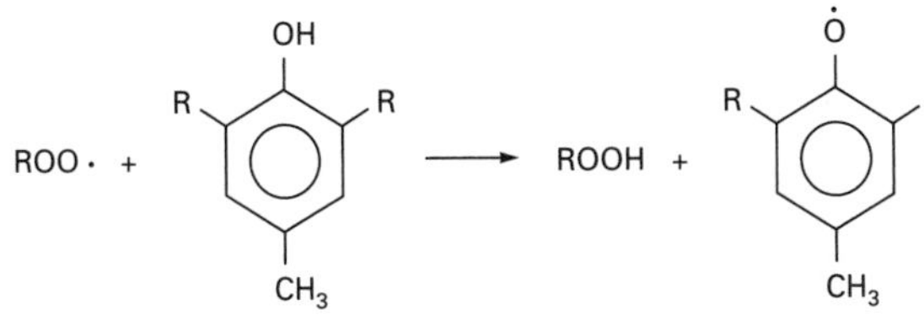

As shown by the following equation, the phenoxy free radical is a resonance species which may couple with a peroxy free radical to produce a stable compound:

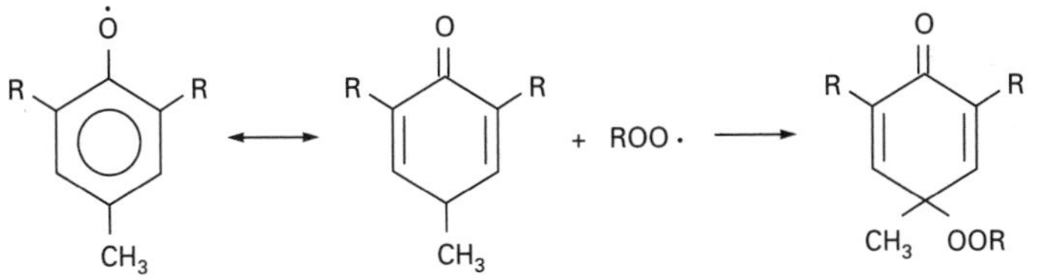

Dialkyl thiopropionates and phosphites are weak antioxidants, but synergistic mixtures, with hindered phenols, are highly effective peroxide decomposers. It is believed that these esters react with hydroperoxides to produce compounds which are able to destroy several equivalents of hydroperoxides. The hydroperoxides are reduced to alcohols by organic phosphites.

Because metal ions are oxidation catalysts, metal deactivators, which chelate with these ions, may be added as part of the stabilization formulation. The catalyst requires association with the hydroperoxide, which is not possible in the presence of metal deactivators, such as oxalyl *bis*-(benzylidenehydrazide) (($C_6H_5$–CH=N–NH–CO–$)_2$).

The use of a metal deactivator in conjunction with a hindered phenol is essential for telephone wire insulation. This use is also advantageous with pigments and fillers that may contain traces of catalytically active metal ions.

## ANTIMICROBIALS

Because natural polymers are subject to attack by biological agents, in the past attempts were made by historians and religious leaders to preserve important books. For example, cedar oil was used by Joshua in order to preserve the books of Pentateuch a few thousand years ago. More recently, it has been recognized that some synthetic polymers or additives, such as plasticizers, in these macromolecules are also subject to enzymatic or chemical attack.

The term "antimicrobials" or "biocides" includes algicides, bactericides, bacteriostats, mildewcides, fungicides, fungistats, microbiocides, germicides, and preservatives. Prior to the 1930s, the principal biocide was the Bordeaux mixture, i.e., slaked limed ($Ca(OH)_2$) in a copper sulfate ($CuSO_4$) solution. Fortunately, the fungicidal activity of dithiocarbamates and the insecticidal properties of phenothiazine ($C_{12}H_9NS$) were discovered in the early 1930s. Some additives, such as organometallic stabilizers, also serve as fungicides.

The principal biocides are copper 8-hydroxyquinolinolate, $N$(trichloromethyl) thio-4-cyclohexane-1, 2-dicarboximide (Captan), di(trichloromethyl) phthalamide (Phaltan), and 2-$n$-octyl-4-isothiazolin-3-one. These additives prevent the discoloration of plastic surfaces resulting from the growth of microorganisms.

Aliphatic ester plasticizers are more susceptible to microbial attack than the aromatic esters, such as dioctyl phthalate. More than 3000 tons of antimicrobials are used annually by the American plastics industry.

Industrial antimicrobial agents are classified as pesticides under the U.S. Federal Insecticide, Fungicide, and Rodenticide Act, and their use is regulated by the Environmental Protection Agency (EPA). One of the newest biocides is 2-methyl-1-naphthyl maleimide.

The American Society for Testing and Materials (ASTM) has developed several tests for the performance of antimicrobials and a guide for mildewcides has been issued under the U.S. Federal Insecticide, Fungicide, and Rodenticide Act by EPA. Biocides are being used in vinyl wall coverings, ditch liners, cable insulation, and single-ply roofing. Antimicrobials are also used in polyurethane rug underlays and carpet cushions.

## ANTIOZONANTS

The reaction of ozone ($O_3$) (ozonolysis) with the ethylenic groups in natural rubber was used by Harries early in the 20th century to show the presence of a 1,4 structure in natural rubber and gutta percha. Unfortunately, the photolysis of oxygen in sunlight in urban areas has generated relatively high concentrations of ozone (5 to 10 ppm), which degrades rubber tires as a result of chain scission of the polyisoprene molecule.

This cracking of stretched polydiene elastomers may be prevented by the addition of paraffin waxes, inert polymers, or antiozonants, such as *N,N*-substituted *p*-phenylenediamine (PDA). Ozone attack on olefinic double bonds in elastomers is deterred by a wax which blooms to the surface or by a scavenger that reacts preferentially with ozone. The major commercial antiozonants are *N,N'*-di(2-octyl)*p*-phenylenediamine (($C_6H_{12}C_2H_5(NH)_2C_6H_4$), or Antozite) and Antozite 67, methylheptyl (Antozite 2), and the methylbutyl, phenyl derivatives of PDA. Other PDA-based antiozonants are available from Goodyear and Naugatuck Chemical under the trade names of Wingstay and Flexzone. More than 30,000 tons of substituted PDAs are produced annually in the U.S.

## ANTISAG AGENTS

Antisag or thickening agents increase the viscosity of or serve as thixotropes for prepolymers or high solids polymeric solutions. Thixotropes thicken these polymeric systems, but lose this property when stirred. Fumed silica, bentonite clay, and hydrated silicate minerals have been used as thickening agents.

## ANTISLIP AGENTS

Antislip agents are similar to antiblocking agents. These agents, such as oleamide or erucamide, may be incorporated in the polymer (usually polyolefins) or sprayed on the surface.

## ANTISTATS

The existence of static electricity was observed in 600 B.C. by Thales of Miletus, who demonstrated that rubbed amber attracted dust. A triboelectric series, which showed the degree of electrostatic susceptibility of commonly available materials, was proposed in 1757 by Wilcke. Wool and glass were rated above rubber in this series.

Most polymers are poor conductors of electricity and tend to build up electrostatic charges. Electrostatic discharge may result in an electrical shock or even cause explosions and fires. This problem has been solved, to some extent, by the addition of conductive substances (antistats), such as tetracyanoquinodimethane (TCNQ), metal fibers or flakes, or acetylene black, or by copolymerization with more polar reactants.

Fatty quaternary ammonium compounds, amines, imidazolines, ethoxylated fatty amines, phosphate esters, polyethylene glycols, and polyethylene glycol fatty acid esters are used as antistats. It is assumed that these additives migrate to the polymer surface and form a continuous thin conductive layer. Typical unmodified polymer resistivities are of the order of $10^{16}$ to $10^{18}$ $\Omega$, and these values can be reduced to $10^{10}$ to $10^{11}$ $\Omega$ by the addition of 0.1% of an antistat.

A good balance between an antistat's compatibility and diffusibility is essential for obtaining useful charge decay rates. The time for the charge to reach one-fourth of its original value may be determined from the surface resistivity and should be less than one second. The half-life period ($t_{1/2}$), at which the voltage is one-half that of the original voltage, is also proportional to the surface resistance and the capacitance of the polymer. There is a reciprocal relationship between the surface resistivity and the relative humidity. In the absence of an antistat, low values of humidity, of the order of 20%, prevent charge equalization over the surface of the polymer.

## BLOWING AGENTS

A blowing or foaming agent is a substance that releases gas when heated to a specific temperature. A classical blowing agent, sodium bicarbonate, which is also used in cakemaking, has been augmented by azo compounds, such as azobisisobutyronitrile (AIBN), nitroso compounds, such as *N,N*-dimethyl-*N,N*-dinitrosoterephthaleimide, and sulfonyl hydrazides, such as benzene sulfonyl hydrazide. Citric acid and sodium

| Table 2.3  Annual U.S. Use of Chemical Blowing Agents | |
|---|---|
| **CBA** | **Consumption, tons** |
| Azodicarbonamide (1,1-azobisformamide, ABFA) | 43,000 |
| Modified azodicarbonamides | 8,300 |
| Oxybisbenzene sulfonyl hydrazide (OBSH) | 2,300 |
| High-temperature CBA's | 2,300 |
| Citric acid-sodium bicarbonate | 1,600 |
| **TOTAL** | **57,500** |

bicarbonate, which have been used as blowing agents in cakemaking and in polymeric foams, are also used as nucleating agents.

Some organic chemical blowing agents (CBA) release the propellant gas over narrow temperature ranges. Hence, these CBA's are classified as low-, medium-, and high-temperature agents. 4,4'-oxybis(benzenesulfonyl-hydrazide (OBSH) is suitable for foaming at 150 to 160 °C, which is below the normal decomposition temperature of 165 °C. This OBSH evolves 125 $cm^3$/g of nitrogen and a small amount of water vapor.

Most blowing agents are used in the medium-temperature range of 160 to 230 °C. Azodicarbonamide (1,1'-azobisformamide, or ABFA), which decomposes at 205 °C, is the most widely used CBA. As shown in Table 2.3, azodicarbonamides account for more than 90% of all CBA's used annually by the American foamed plastics industry. Many ABFA-type CBA's contain proprietary metal salts that may decrease the decomposition temperature to as low as 150 °C.

High-temperature CBA's, such as *p*-toluenesulfonyl semicarbazide (TSSC), decompose at temperatures above 335 °C. Other high-temperature CBA's such as 5-phenyltetrazole and its analogs decompose in the 335 to 385 °C range.

The decomposition of ABFA is complex, and the gaseous products vary with temperature and the kickers used. However, the following equations are illustrative of its decomposition in the presence and absence of alkalies:

$$H_2N\text{-}CO\text{-}N=N\text{-}CO\text{-}NH_2 \rightarrow N_2 + CO + H_2NCONH_2 \rightarrow HNCO + NH_3$$
$$H_2N\text{-}CO\text{-}N=N\text{-}CO\text{-}NH_2 + 2NaOH \rightarrow NaOOC\text{-}N=N\text{-}COONa + 2NH_3$$
$$NaOOC\text{-}N=N\text{-}COONa + 2H_2O \rightarrow N_2H_4 + 2Na_2\,CO_3 + 2CO_2 + N_2$$

Since sodium borohydride (SBH) ($NaBH_4$) produces much more gas than ABFA and its decomposition is independent of temperature, it is also used as a CBA. The rate of decomposition of SBH may be modified by the addition of heavy metals and by controlling the rate of addition of water. The equation for the decomposition of SBH is:

$$NaBH_4 + 2H_2O \rightarrow NaBO_2 + 4H_2$$

Volatile liquids, such as *n*-pentane, are used as physical blowing agents (PBA). Aliphatic hydrocarbons such as pentane are used for foaming polystyrene (Styrofoam). The chlorine-containing fluorocarbons (CFC) are excellent PBA's, but their use has been discouraged because of the ozone layer controversy. Most nations that have signed the Montreal Protocol have pledged to reduce the production and use of CFC's.

The numbering system for CFC's, which is based on the number of C, H, and F atoms present, was devised by DuPont chemists in the 1930s. These compounds are named as follows: CFC-11 ($CCl_3F$), CFC-12 ($CCl_2F_2$), HCFC-123 ($CF_3CHCl_2$), HCFC-141b ($CCl_2FCH_3$), and HCFC-142b ($CClF_2CH_3$). The atmospheric lifetimes of several CFC's is as follows: CFC-11, 65 years; HCFC-141b, 8 years; and HCFC-141b, 1.5 years.

Because the ozone layer (ozonosphere), which shields the earth from excessive solar radiation, is affected by CFC's, foam producers and environmentalist are concerned about exhausting long-lived CFC's into the atmosphere. Under normal circumstances, the rates of formation of the ozone and the consumption of the ozone in the ozonosphere are equal, but some free radicals produced by the dissociation of some CFC's decrease the concentration of ozone and hence permit an increase in UV radiation on earth.

Amoco Foam Products has phased out its use of CFC's in favor of Amoroam-patented extruded polystyrene foam insulation, which uses HCFC-142b. The European Community environment ministers are backing the phase out of CFC production, and ICI, Montefluor, and ISC chemicals have expanded their production of HCFC-22. DuPont began producing HFC-134a and HCFC-123 in 1990.

ICI plans to produce HCF-134a on a large scale in 1993. Allied-Signal plans to build a facility at Geismar, LA for the production of HCFC-141b. It is estimated that CFC demand in the year 2000 will be as follows: HCFC's, 69%; nonfluorocarbons, 31%; and recycled, 24%.

Methylene chloride has been used as a blowing agent for polyurethane foams, but its use is being discouraged by regulatory agencies.

Physical blowing agents and their boiling points are as follows: *n*-pentane, 36 °C; *n*-heptane, 93 °C; methylene chloride, 40 °C; trichlorofluoromethane, 24 °C; trichlorofluoroethane, 48 °C; isopropyl ether, 68 °C; and methyl ethyl ketone, 80 °C.

Small amounts of gas-forming agents may be used in extrusion of molding compounds to eliminate surface indentations or "sinks" in extrudates or molded products. Larger quantities of blowing agents are used in structural forms and cellular products. Bubbles or cells are formed in polymers when the rate of gas release is greater than the rate of diffusion of the gas. Hence, good foam production requires that the rate of diffusion be less than the rate of gas formation.

# CATALYSIS
## (CURING AGENTS, PEROXIDES)

Many of the so-called curing agents used by the polymer industry do not meet the definition of catalysts, since they are reactants. Initiators, such as benzoyl peroxide (BPO), used to initiate the polymerization of vinyl monomers or to crosslink linear polymers actually react irreversibly with other reactants to produce radicals which undergo further reaction. Accelerators used to enhance the reaction of sulfur and rubber also react irreversibly. Likewise, substituted amines and cyclic anhydrides used as curing agents for epoxy resins react irreversibly to produce network polymers. Urethane catalysts such as amines and organotin compounds are also included in this section.

Strong acids and hexamethylenetetramine were used by Baekeland to cure phenolic resole and novolac resins, respectively. These additives are still in use today. Strong acids are also used to cure urea and melamine resins and to cationically initiate the polymerization of monomers, such as isobutylene and furfuryl alcohol.

The most widely used additives employed for initiating polymerization reactions are organic peroxides, such as BPO, which is used at an annual rate of 4000 tons in the U.S. Other major organic peroxy initiators are methyl ethyl ketone peroxide (MEKP), peresters, and dialkyl peroxides, such as dicumyl peroxide, which are used at an annual rate of 4600, 3400, 900, and 2000 tons, respectively. Lauroyl and decanoyl peroxides are used at an annual rate of 625 tons and 125 tons, respectively.

The list of perester initiators includes *t.*-alkyl peroxyesters (such as *t.*butyl, *t.*amyl, and *t.*octylperoxy-2-ethylhexanoate), neohexanoate, and neodecanoate. Alkyl peroxydicarbonates and perketals, such as *n*-butyl-4.4-di(*t.*-butylperoxyvalerate), are also being used. AIBN, which decomposes to form nitrogen and a free radical, is also used as an initiator. Transition metal catalysts, such as ruthenium compounds, can be used to initiate the polymerization of monomers in aqueous systems.

Activators or promoters, such as *N,N*-dimethylaniline, are used with organic peroxides such as BPO to cure polyester prepolymers at ordinary temperatures. Ultraviolet radiation is also used in the presence of photosensitizers, such as benzoin ether, and a reducing agent, such as diethanolamine, to cure polyester prepolymers.

More than 2400 tons of amine or organotin catalysts are consumed annually by the U.S. polyurethane (PUR) industry. Tertiary amines, such as triethylenediamine (DABCO) and *N*-ethylmorpholine, are used to catalyze the reaction of the isocyanate group with water. Organotin compounds, such as tin octoate, which are sometimes called gelation catalysts, are used to catalyze the reaction of isocyanate groups with hydroxyl groups. The amine catalyst aids the production of carbon dioxide gas. Controlled amounts of tin catalysts modify the rate of polymerization so that strong unicellular foams are produced.

Potassium salts of organic acids, such as potassium octoate, and ammonium salts, such as 2,4,6-*tris*(*N,N*-dimethylaminomethyl) phenol, are used as catalysts for the production of isocyanurate foams. Experimental design, statistical analysis, and computer modeling are being used in the development of improved PUR catalyst systems.

Accelerators are used as catalysts for the curing of rubber with sulfur (vulcanization). The widely used accelerators are derivatives of 2-mercaptobenzothiazole (Captax), which was introduced in the early 1920s. Ultraaccelerators, which catalyze the crosslinking of rubber with sulfur, are derivatives of thiocarbamates and thiuram sulfides.

Activators such as zinc oxide and stearic acid are usually added to rubber compounding recipes to speed up the crosslinking reaction. Synthetic polymers of butadiene are vulcanizable by sulfur, and those containing active functional groups, such as neoprene and sulfochlorinated polyethylene, may be crosslinked by the addition of heavy metal salts, such as litharge.

Copolymers of ethylene and vinyl acetate, copolymers of ethylene and propylene (EPM), unsaturated polyester, and silicones, which are not curable by sulfur, are crosslinked by organic peroxides, such as dicumyl peroxide. The latter and *bis*(*t*.butyl peroxy isopropyl)benzene are used as curing agents for olefinic polymers. *Bis*(2,4-dichlorobenzoyl) peroxide is used for curing silicones and 2,5-dimethyl-2,5-di(*t*.butylperoxy) hexyne-3 is used for curing high-density polyethylene (HDPE).

It is possible to obtain polymers with almost any desired degree of crosslinking (crosslink density) by controlling the amount of curing agent used. *t*.Amyl peroxide has been used for crosslinking vinyl polymers. *t*.Butyl peroxybenzoate is used for high-temperature reactions (150 °C), and 1,1-di(*t*.butylperoxy) cyclohexane is useful for lower temperature applications.

*t*.Butylperoxy neodecanoate exhibits greater reactivity at elevated temperatures and high pressure than other peroxy catalysts used for the production of low-density polyethylene (LDPE), but acetyl 2-chloroocta-noyl peroxide is preferred for the production of PVC and 2,5-di(ethyl-hexanoyl peroxy) hexane is preferred for the polymerization of *p*-methyl-styrene at 100 °C.

Photosensitizers, such as benzil and benzoin ether, and amine reducing agents, such as diethanolamine and benzyldimethylamine, are used in initiator systems that are activated by specific wavelengths of light. Electron beam crosslinking is also being used for curing sensitized polymer systems.

Crosslinking with peroxy compounds is enhanced by the use of co-agents, such as triallylcyanurate. The annual U.S. consumption of peroxy compounds is shown in Table 2.4. Unsaturated hydrocarbon polymers have been crosslinked by the addition of phenolic resins and acid catalysts, *p*-dinitrosobenzene and lead peroxide, and alkoxysilanes and moisture.

Most epoxy resins are based on the reaction of bisphenol-A and epichlorohydrin. Thus, diglycidyl ether (EPON, Araldite) may be crosslinked

| Table 2.4  Annual U.S. Use of Peroxy Compounds | |
| --- | --- |
| **Compound** | **Consumption, tons** |
| Benzoyl peroxide | 4,590 |
| Decanoyl peroxide | 1,360 |
| Lauroyl peroxide | 636 |
| Dialkyl peroxides | 2,318 |
| MEK peroxide | 4,864 |
| Peresters | 3,954 |
| **TOTAL** | **18,045** |

(cured) at room temperature in the presence of a trace of water and polyamines, such as diethylenetriamine or amides with amine end groups (Versamids). Epoxy resins may also be cured by heating with a cyclic anhydride, such as phthalic anhydride.

The amines react with the terminal epoxy groups by a typical oxirane-amine reaction mechanism. The anhydrides react with the more abundant hydroxy pendant groups to produce half esters. The carboxyl groups produced then react with other hydroxyl groups.

## COLORANTS

The first colorants used in plastics and rubber were inorganic pigments, such as the iron oxides (ochers, umbers, and siennas) and carbon black. These classic pigments have been augmented by the introduction or organic pigments, such as phthalocyanines and dyes. More than 200 different colorants are available commercially. More than 350,000 tons of pigment concentrates and dry pigments are used annually by the American plastics industry. Small amounts of encapsulants, paste dispersions, and liquid colors are also used. Inorganic pigments account for more than 125,000 tons annually in the U.S., but phthalocyanines account for less than 5000 tons.

While titanium dioxide and carbon black (see Chapter 3) are the principal colorants used by the polymer industry, both natural and synthetic iron oxides and many organic pigments and dyes are also used. Dyes are complex organic molecules containing at least one chromophore or color-forming group. They are usually completely soluble, but pigments are insoluble in the polymer.

The principal dyes are azo ($-N=N-$) dyes, such as acid yellow 36; anthraquinone dyes, such as solvent red 111; xanthene dyes, such as basic violet 10; and azure dyes, such as indulene. The principal organic pigments are monazo pigments, such as permanent yellow FGL; diazo pigments, such as benzidine yellow; quinacridone pigments, such as pigment violet 19; diaxazines, such as carbazile dioxazine violet; vat pigments, such as isoviolanthrone violet; perylene pigments, such as PR 123; thioindigo pigments, such as pigment red 88; phthalocyanine pigments, such as phthalocyanine blue; and tetrachloroisoindolinones.

The classical red pigments are acid monazo reds, anthraquinone reds, perylene, anthraquinone and diazo condensation reds, and thioindigo. The principal blue and green pigments are the phthalocyanine pigments. Special effects are obtained by the addition of fluorescent and pearlescent pigments. Inorganic pigments are usually metallic oxides that have been calcined at high temperature. Most pigments have an average particle size of 0.01 to 1.0 µm. The inorganic pigments include natural red and synthetic iron oxides, yellow, maroon, cadmium, and mercadmium pigments. Chromium oxide, cobalt blue, ultramarine blue, and cobalt blue are used as blue and green pigments. Soluble blue and green colorants are also used.

Synthetic iron oxides, carbon black, and nigrosine are used as black colorants. Special effects are provided by aluminum and bronze powder. Titanium dioxide and zinc oxide serve as white pigments and improve the weatherability of plastics. Zinc sulfide and white lead are also white pigments, but are not used to any great extent in plastics.

Reactive colorants can be used to measure the crosslink density of polymers. Special effects are obtained by the use of metallic, pearlescent, and fluorescent pigments. Despite long-time speculation about health hazards, cadmium pigments and colorants containing traces of chlorinated biphenyl (PCB) continue to be used to a limited extent.

The annual consumption of pigments and dyes by the U.S. plastics industry is shown in Table 2.5.

| Table 2.5  Annual U.S. Use of Pigments and Dyes | |
| --- | --- |
| Colorant | Consumption, tons |
| Iron oxide | 4,360 |
| Cadmium pigments | 2,360 |
| Chrome yellow | 2,370 |
| Molybdate orange pigment | 1,660 |
| Phthalo-blues | 1,400 |
| Phthalo-greens | 560 |
| Organo reds | 1,150 |
| Organo yellows | 210 |
| Nigrosine | 1,450 |
| Water-soluble dyes | 220 |
| Anthraquinones | 220 |
| **TOTAL** | **20,000** |

## CONDUCTIVITY AIDS

While most polymers are inherently nonconductive, there is a need for some electrical conductivity to avoid electromagnetic interference (EMI) in office machine housings, etc. Some polymers, such as polyacetylene, are fair conductors; this quality can be enhanced by the addition of electron donor or electron acceptor dopants.

The conductivity of plastics can be enhanced by the addition of as little as 1% of very fine steel wire, aluminum flakes treated with organosilanes, or metal-coated fiberglass. Thirty-five thousand tons of plastics are used in business machines, but less than half of this volume meets EMI specifications.

Arsenic pentafluoride ($ASF_5$), which acts as an electron acceptor, is used as a dopant for $p$-type polyacetylenes. Lithium salts are also used as electron donor dopants to produce $n$-type polyacetylenes. Typical $p$-type composites have conductivities as high as $10^3 \ \Omega \cdot cm^{-1}$. Typical $n$-type composites have conductivities as high as $200 \ \Omega \cdot cm^{-1}$. Iodine is used as the additive in poly-2-vinylpyridine, used in pacemaker batteries. The addition of conductors will also convert nonconductive unsaturated polymers, such as natural rubber, to conductors.

## COUPLING AGENTS

While the bond between carbon black and natural rubber is satisfactory, the bond strength can be improved by the addition of coupling agents with functional groups which are attached to the surface of the filler and the rubber. The use of coupling agents is not widespread in the elastomeric industry, but these additives do reduce the time of mixing between fillers and some elastomers, such as butyl rubber.

The pioneer coupling agent, called a promoter, was $p$-nitrosodimethylaniline ($NOC_6H_4N(CH_3)_2$). It is still available under the trade name of Elastopar. Knowledge of the importance of diverse functional groups which would be attracted by the filler and polymer was useful for the more vital development of coupling agents for forming molecular bridges at the interfaces between glass fibers and polyester resins in the early 1940s.

The first commercial coupling agent used in fibrous glass-polymer composites was a Werner-type chrome complex (methacrylato chromic chloride, or Volan). The Volan coupling agent has been replaced, to a large extent, by organosilanes, organotitanates, organozirconates, stearates, chlorinated paraffins, carboxylic acids, polymerizable cellulosics, and wetting agents. These coupling agents also upgrade the properties of filled polymers.

It has been suggested that these agents modify the interfacial region between the filler and polymer and provide an improved interfacial bond that permits higher loadings and upgrades the physical properties of the composite. Wetting agents wet the filler and ensure a homogeneous dispersion of the wetted particles in the resin. The hydrophobic wetting agents also displace air, contaminants, and moisture on the hydrophilic filler surface.

The chemical structure of silane coupling agents can be represented by the general formula $(RO)_3SiR'X$, in which X represents a functional organic group, such as an amine, methacrylate, epoxy, etc. The functional group is attached to the silicon atom through a stable hydrocarbon linkage ($R'$). The RO groups are alkoxy or acetoxy groups which, in the presence of moisture, are hydrolyzed to silanol ($Si(OH)_3$) groups which are attracted to the filler surface.

The organotitanate and organozirconate coupling agents react with free protons at the filler surface interface to form a monomolecular layer that is compatible with the polymer. Since these coupling agents reduce the viscosity of the resin-filler mix, high loadings are possible. For example, a 90% loading of ferric oxide in nylon-66 exhibits virgin flow. In general, the production energy is decreased and the production rate is increased when small amounts of titanate coupling agents are present in the resin-filled composite.

According to Plueddemann, the original titanates were not true coupling agents, since they do not provide water-resistant bonds. Both ortho-silicates and titanates are surface-active materials that improve processing and provide adhesion to filler surfaces through the formation of an inter-penetrating network. About 4000 tons of silane coupling agents are used annually by the U.S. polymer industry. It has been demonstrated that as little as 0.2% vinylsilane converts silica into a reinforcing filler.

The effectiveness of calcium carbonate as a semireinforcing filler has been improved by the addition of ethylene-ethyl acrylate copolymers. A polymerizable acrylmethyl cellulose ester is being used as a coupling agent for fillers in bulk and sheet molding compounds. Maleated propylene waxes have been used as coupling agents in filled polypropylene, and the impact and stiffness qualities of filled polyolefins have been improved by the use of surface-modified talc, mica, and alumina tri-hydrate (ATH) fillers.

Organosilanes are not effective coupling agents for carbon or aramid fibers, but they are extremely useful as surface modifiers for fibrous glass in polymer composites. There are now four major suppliers of surface-treating agents. The plastics and rubber industries use 40 and 30% of this volume, respectively, and the remainder is divided between the adhesive-sealant and ink-coatings industries. Seven thousand tons of coupling agents were produced in the U.S. in 1989, and it is anticipated that this volume will increase to 10,000 tons in 1993.

Because silanes are not effective as coupling agents for filled poly-olefins, other additives such as zircoaluminum organic compounds and silanosulfonyl azide have been used. Production of this azide additive was abandoned a few years ago, but it is now again available. The concentration of the carbonyl bond has been used to monitor the rate of reaction of silane to glass surfaces. The total annual volume of nonsilane coupling agents is about 3000 tons; the titanates account for 75% of this volume.

A dramatic increase in viscosity is observed when neoalkoxytitanate or zirconate coupling agents are added to PVC. Superior flexural strength properties of composites are also observed when mineral fillers are pre-treated with organosilanes.

Improvements in the impact and stiffness of filled polyolefins is noted when surface-modified talc is used as a filler. Maleated polypropylene waxes have been used successfully as coupling agents in filled polypropy-lene. The impact resistance of clay-filled nylon is improved by use of surface-modified kaolin, such as chlorokaolin.

## FLAME RETARDANTS

Most organic polymers are thermally unstable and form volatile decomposition products at elevated temperatures. Because the residue is usually porous, it facilitates oxygen penetration and additional pyrolysis of the residue. It is believed that combustion is a free radical chain reac-tion in which the rate is related to ignition susceptibility and rate of for-mation of hydroxyl free radicals ($\bullet$OH), as shown in the following equa-tions for this self-sustaining exothermic reaction:

$$\overset{|}{\underset{|}{C}}H_2 \overset{\Delta}{\to} \bullet\overset{|}{\underset{|}{C}}H \overset{O_2}{\to} \overset{|}{\underset{O_2}{C}}HOO\bullet \to \overset{|}{C}HO + \bullet OH$$

$$H\bullet \to HOO\bullet + HO\bullet + O\bullet \overset{H_2}{\to} H_2O + HO\bullet$$

$$H\bullet \overset{O_2}{\to} H_2O$$

About \$30 billion is lost and about 6000 people die annually in the U.S. as a result of fire. Fire potential is increasing because of the prox-imity of buildings and the increase in arson in urban areas. History has recorded many disastrous fires, such as those in Chicago, London, Mos-cow, Rome, San Francisco, and Tokyo. However, most fires and fire inju-ries occur in one- and two-family homes. The U.S. and Canada lead the world in death rate from fires.

Fortunately, the evolution from manual water pumping to horse-drawn steam pumpers to modern fire engines has reduced the rate of increase in conflagrations, and the more recent use of fog and spray nozzles and wetting agents, such as fluorocarbon surfactants (light water), has reduced the extent of fire damage. It is of interest to note that 75% of all U.S. homes have smoke detectors; however, other countries prefer to stress preparedness over detectors.

Additional reduction in the frequency of fires and fire damage may be anticipated as a result of the addition of flame retardants to building materials and apparel. Wood and paper products, impregnated with ammonium salts such as ammonium sulfamate ($NH_4OSO_2NH_2$) and borax ($Na_2B_4O_7 \cdot 10H_2O$) or boric acid ($H_3BO_3$), produce fewer volatile products when heated, and hence are more flame resistant.

Many other flame retardants, such as alumina trihydrate ($Al_2O_3 \cdot 3H_2O$), antimony oxide ($Sb_4O_6$), brominated and chlorinated hydrocarbons, phosphorus-containing polyols, and halogenated phosphate esters, are used as flame retardants for many polymers.

Alumina trihydrate (Gibbsite) is an intermediate in the production of aluminum, which is produced from bauxite. ATH, which is a nontoxic filler that releases water when heated, has a Mohs' hardness of 3 and is available as a surface-treated product. The major peak for ATH on a DTA curve at 312 °C represents endothermic dehydration to the monohydrate and release of water. ATH retards pyrolysis, reduces the burning rate, and acts as a smoke suppressant. Loadings as high as 70% are possible when ATH is treated with an organosilane coupling agent.

Antimony oxide reacts synergistically with halogen compounds. Because antimony oxide does not volatilize at flame temperatures, halogen compounds must be present to ensure the formation of the more volatile antimony halides. The trioxide is the most widely used antimony oxide flame retardant.

The flame-retardant efficiency of organic chlorides is in the order of aromatic, alicyclic, and aliphatic, which is the reverse of thermal efficiency. The relative activity of halogenated paraffins as flame retardants is as follows: tertiary is more active than secondary, which is more active than the primary halogen compounds. Despite their limitations, halogenated paraffins are the most economical flame retardants. Polypentabromobenzyl acrylate has been used as a flame retardant for nylon 6 and polybutylene terephthalate (PBT). Tetrabromophthalate diol, which forms considerable char, has been used as a flame retardant for polyurethane foams.

The brominated hydrocarbons are more efficient flame retardants, but they may tend to bloom to the surface and decrease the ultraviolet light stability. Since its heat stability is in excess of 129 °C, brominated polystyrene is being used as a flame retardant in nylon, polyethylene terephthalate (PET), and PBT. While the U.S. Environmental Protection Agency has not regulated halogen-containing flame retardants, Germany has requested that the European Economic Community restrict the use of polybrominated diphenyl oxides (PBDPO) used in composites. PBDPO also appears to pose a workplace health hazard. General Electric has discontinued the use of PBDPO in its PBT composites, and Mobay, G.E., and Akzo are promoting the use of halogen-free flame retardants. Claims that 2,3,7,8-tetrabromodioxane and furan are toxic have been questioned by investigators at Triangle Laboratories, who showed that dioxanes are produced only in small amounts (10 ppb) when the compounded plastics are heated for long periods of time at elevated temperatures under conditions that are highly unlikely in a "real world" situation. Studies have also shown that PBDPO is noncarcinogenic.

The production of flame retardants in the U.S. in 1989 and that projected for 1994 is shown in Table 2.6.

While the mechanism of flame-retardant phosphorus compounds is not well understood, it is assumed that they enhance the thermal decomposition of polymers and that phosphoric acid ($H_3PO_4$) forms a protective char that reduces afterglow on the surface. Halogenated organophosphorus flame retardants have been used in polyurethane foams.

Zinc borate ($(Zn)_3(BO_3)_2$) has been used as a partial replacement for antimony oxide. Zinc borate is being used in combination with ATH as an effective flame retardant and smoke suppressant in EVA, HDPE, EDPM, EEA, epoxies, and acrylics. Molybdenum compounds, such as molybdenum trioxide ($MoO_3$) and ammonium molybdate ($(NH_4)_6Mo_7O_{24} \cdot 4H_2O$) are used as flame-retardant smoke suppressants.

Internal flame retardants (reactive flame retardants), such as chlorendic anhydride ($C_9H_2Cl_6O_3$), tetrabromobisphenol-A ($(HOC_6H_2Br_2)_2(C(CH_3)_2)$), and phosphorus-containing polyols such as glycerylphosphoric acid ($C_3H_5(OH)_2H_2PO_4$) are used to produce flame-retardant polyesters, polycarbonates, and polyurethanes. Diallyl tetrabromophthalate has also been used as a flame-retardant monomer.

Some complex intumescent additives, such as starch, sorbitol, or pentaerythritol, and blowing agents, such as melamine or urea and phosphates,

| Table 2.6  U.S. Production of Flame Retardants | | |
|---|---|---|
| | Production, tons | |
| Type | 1989 | 1994 |
| Alumina trihydrate | 144,000 | 198,000 |
| Antimony oxides | 23,000 | 29,000 |
| Bromine compounds | 27,000 | 36,000 |
| Chlorine compounds | 41,000 | 56,000 |
| Phosphorus compounds | 87,000 | 110,000 |
| Others | 21,000 | 27,000 |
| **TOTAL** | **344,000** | **456,000** |

phates, yield incombustible gases that produce surface foam and expel oxygen. Other additives, such as ammonium sulfate or ammonium sulfamate, produce incombustible products when heated.

Tests for flame retardancy, including the limiting oxygen index (LOI), Underwriters Laboratory (UL) tests, and National Bureau of Standards (NBS) smoke box tests, have been used in an attempt to correlate the results of these tests on PVC, unsaturated polyesters, high-impact polystyrene (HIPS), ABS, HDPE, and polypropylene (PP). In the LOI tests, the concentration of oxygen, in a mixture of oxygen and nitrogen, at which a candle-like specimen continues to burn is called the LOI value (ASTM D-2863-77). Since it is alleged that most fire deaths result from the inhalation of toxic fumes, attempts are being made to develop more realistic toxicity tests.

A silicon-based flame retardant that reacts synergistically with magnesium stearate, in the presence of a small amount of decabromodiphenyl oxide and ATH, has been described. The tracking of flame-resistant nylon has been reduced by the addition of iron and zinc oxides.

More than 340,000 tons of flame retardants are used annually in the U.S. The annual consumption of specific flame retardants is shown in Table 2.7.

| Table 2.7  Annual U.S. Use of Flame Retardants | |
|---|---|
| **Type** | **Consumption, tons** |
| **External additives** | |
| Alumina trihydrate | 144,000 |
| Antimony oxides | 23,000 |
| Boron compounds | 5,000 |
| Chlorinated aliphatics | 41,000 |
| Phosphate esters | 87,000 |
| **Internal additives** | |
| Epoxy intermediates | 10,000 |
| Polycarbonate intermediates | 3,000 |
| Polyester intermediates | 6,000 |
| **TOTAL** | **319,000** |

## HEAT STABILIZERS

Polymers with tertiary hydrogen atoms, such as polypropylene, low-density polyethylene, polystyrene, polyvinyl chloride, and most other vinyl polymers, are thermally unstable. Thus, in the absence of stabilizers, these polymers tend to decompose at processing temperatures and produce the peroxy radicals described in the section "Antioxidants" in this chapter. With the exception of PVC, these polymers can be heat stabilized by the addition of antioxidants.

In contrast, because PVC requires additional heat stabilization, it was not commercialized until the early 1930s. Even then, the product was a plasticized PVC (Koroseal) with a much lower processing temperature than rigid PVC. Degradation of rigid PVC starts at 100 °C.

As shown by the following equation, the allylic chlorine atoms, which are usually present in PVC, are readily abstracted and hydrogen chloride (HCl) is produced. Hence, inorganic basic lead compounds, such as basic lead carbonate ($2PbCO_3 \cdot Pb(OH)_2$) were the classic stabilizers for PVC.

$$-CH_2-CHCl-CH_2-CHCl-CH=CH- \xrightarrow{\Delta}$$
$$-CH_2-CHCl-CH=CH-CH=CH + HCl$$

This product, which has an allylic chlorine atom, also loses hydrogen chloride to produce a highly colored conjugated unsaturated structure, as shown by the following equation:

$$-CH_2-CHCl-CH=CH-CH=CH- \rightarrow -CH=CH-CH=CH-CH=CH-$$

As shown by the following equation, the formation of the conjugated unsaturated structure is prevented by the presence of heat stabilizers, such as barium or cadmium salts of organic acids:

$$-CH_2-CHCl-CH=CH- \xrightarrow{Cd(OOCR)_2}$$
$$-CH_2-CH(OOCR)-CH=CH- + CdCl_2$$

However, since the product ($CdCl_2$) is a Lewis acid, which can promote additional degradation, the barium salt is used to convert the inorganic salt to a heat-stabilizing salt of an organic acid, as shown by the following equation:

$$CdCl_2 + Ba(OOCR)_2 \rightarrow Cd(OOCR)_2 + BaCl_2$$

Other stabilizer systems containing phosphites, such as triphenyl phosphite, will form desirable chelates with cadmium. Epoxy compounds, such as epoxidized tall oil esters, also serve as plasticizers and react with the allylic chlorine atoms.

Nevertheless, sulfur-containing organotin compounds such as butyltin mercaptides are now recognized as being the most efficient heat stabilizers for PVC. However, they are odoriferous and have poor light stability.

The annual consumption of PVC heat stabilizers by the U.S. plastics industry is shown in Table 2.8.

## IMPACT MODIFIERS

The brittle nature of several intractable plastics, such as cellulose nitrate and PVC, has been overcome by the addition of low molecular

| Table 2.8  Annual U.S. Use of PVC Heat Stabilizers | |
| --- | --- |
| **Stabilizer** | **Consumption, tons** |
| Barium-cadmium compounds | 13,600 |
| Tin compounds | 11,100 |
| Lead compounds | 9,700 |
| Calcium-zinc compounds | 2,300 |
| Antimony compounds | 1,100 |
| **TOTAL** | **37,800** |

weight compounds (plasticizers), such as camphor and dioctyl phthalate. However, Celluloid is no longer a large-volume plastic, and while plasticized PVC (Koroseal) is used in large quantities, there is a need for less flexible products with good impact resistance. In contrast, polystyrene and polyolefins are not readily flexibilized by the addition of liquid plasticizers, and there is a need for styrene and olefin polymers with better impact resistance.

Accordingly, the impact resistance of PVC is improved by the addition of the terpolymer of acrylonitrile ($CH_2$=CHCN), butadiene ($H_2C$=CH–CH=$CH_2$), and styrene ($H_2C$=CH–$C_6H_5$), which is called ABS. The comparable terpolymer of methyl methacrylate ($CH_2$=C($CH_3$)COOC$H_3$) (MBS) or the polymer obtained from acrylonitrile, methyl methacrylate, butadiene, and styrene (AMBS) is used in place of ABS. Blends, in which both phases are rigid, have improved melt flow, reduced shrinkage, and improved mechanical properties. A transparent bottle is produced from a blend of PVC and a methyl methacrylate, butadiene, styrene graft terpolymer (MBS) with a similar index of refraction.

Other impact modifiers for PVC include polyalkyl acrylates, ethylene-vinyl acetate copolymer (EVA), and chlorinated polyethylene (CPE). The latter is preferred in Europe and EVA is favored in the USA. The extent of impact improvement is dependent on the particle size and the concentration of the additive. Tensile strength, hardness, and gloss are reduced as the concentration of the additives is increased. Acrylic-modified PVC is used for siding, windows, gutters, and outdoor furniture.

High-impact polystyrene is produced by the polymerization of styrene in which an elastomer, such as SBR, is dissolved. A phase inversion occurs during polymerization so that the finally dispersed elastomer particles are grafted on polystyrene. Block copolymers of styrene and butadiene also have improved impact resistance.

The rate of production of blow-molded LLDPE may be increased by the addition of small amounts (0.05%) of fluoroelastomers. Block copolymers of styrene and butadiene (Kraton) also improve the dart impact qualities and permit up to 33% downgauging of LLDPE film.

These block copolymers, which are also used as impact modifiers for polybutylene, butyl rubber, and SBR, have also been used as impact modifiers for PP; however, ethylene-propylene elastomers (EPM and EPDM) are the major additives used today. These modified PP's are used for shoe heels, hose, and cable insulation. High-impact polyamides (nylons) have been produced by the addition of EPDM. The thermoformability of polycarbonate is improved by the addition of ABS and PBT. The latter enhances stress-crack resistance. Plasticizers in cellulosics, such as cellulose acetate butyrate (CAB), have been replaced by EVA.

## LOW-PROFILE (LOW-SHRINK) ADDITIVES (LPA)

Reinforced unsaturated polyesters undergo considerable shrinkage during polymerization. While this characteristic aids in the removal of molded parts, it also produces a wavy surface, which is magnified when the surface is painted. This shrinkage, which may also be accompanied by warpage and the formation of internal cracks and voids, can be lessened if the system is converted from a one-phase to a two-phase system.

These systems exhibit a sudden expansion coincidental with the exothermic polymerization, which is followed by a 7% shrinkage. Polyesters with low-profile resin additives do not undergo the shrinkage until the polymerization is complete. Thus, the final volume is about 3% greater than the initial volume. This effect is not noted with hand lay-up techniques because of the absence of the rapid initial exotherm and lack of sudden thermal expansion.

The pioneer LPA was finely divided polyethylene. The use of polymethyl acrylate as an LPA was patented in 1968 and continues to be the standard. Other LPA resins are polystyrene, cellulose acetate butyrate, PVC, polyvinyl acetate, polyurethane, Nevlon (Union Carbide), and polycaprolactone. All mechanical properties of composites are decreased, but impact resistance is increased when LPA's are present.

## LUBRICANTS

Polymers tend to adhere to metal surfaces during processing. This difficulty can be overcome by the addition of small amounts (0.5%) of lubricants, which improve polymer flow, reduce adhesion to metals, and permit extrusion through dies, etc., in the shortest period of time. Low molecular weight hydrocarbon waxes and silicones are used as internal lubricants and higher molecular weight polyethylene and oxidized polyethylene are used as external lubricants. In addition to serving as lubricants, these additives also reduce die swell in the extrusion process.

The most important lubricants are $C_{16}$ to $C_{18}$ fatty acids; stearic acid is the principal additive. Oxystearic acid and stearyl alcohol obtained by the hydrogenation of castor oil, montan wax, calcium stearate, and oxidized polyethylene are also used. Lubricants are essential for processing PVC. Lubricants are also used for processing polystyrene and styrene copolymers, polyolefins, polyamides, polycarbonate, polyacetals, and polyurethanes. More than 4000 tons of lubricants are used annually by the U.S. polymer industry. The relative amounts of these lubricants are as follows: calcium and zinc stearates, 42%; paraffin wax, 28%; fatty acid amides, 18%; and fatty acid esters, 12%.

## MOLD RELEASE AGENTS
## (PARTING AGENTS, RELEASE AGENTS)

Most mold release agents are external additives, such as metallic soaps, which are sprayed on the mold surface. Finely divided solids, such as mica and talc, are applied as external parting agents. Because the application of external mold release agents is labor intensive, there is a tendency to use internal agents, such as stearic acid and silicones, which bloom to the surface. The use of silicones in the reaction injection molding (RIM) of polyurethanes has reduced labor costs and improved the surface of molded parts. About 200 tons of internal mold release agents are used annually by the American plastics industry.

## NUCLEATING AGENTS

Some polymers, such as polystyrene, polycarbonate, and most elastomers, are amorphous. Some, such as PVC, have a small degree of crystallinity, and others, such as the polyolefins and nylons, are highly crystalline. The degree of crystallinity and size of the spherulites in the crystalline polymer are affected by the rate of cooling from the melt and the presence of nucleating agents.

Spontaneous nucleation of a cooled or supercooled melt is usually dependent on uncontrollable concentrations of foreign substances. Since a high degree of crystallinity of crystalline polymers is essential for optimum qualities, such as hardness, tensile strength, and modulus, it is advantageous to add nucleating agents.

The translucency of LDPE film has been reduced for several decades by the addition of finely divided benzoic acid ($C_6H_5COOH$). Studies of the effects of these empirically added insoluble substances has resulted in the introduction of additional nucleating agents. It is now recognized that these additives must have higher melting points and be insoluble in the polymer, yet they must be absorbed by the polymer.

These criteria are readily met by finely divided fillers (such as carbon black, silica, calcium carbonate, talc, or clay), by pigments (such as cadmium red and chromic oxide), by salts of organic acids (such as sodium *p-t*.butyl benzoate), and by some incompatible polymers (such as ionomers and polyolefins). Transparent moldings of nylon-66 are obtained when fumed silica is dispersed in the polymers.

It is also advantageous to add nucleating agents to PET and polyolefins. Potassium stearate is not effective in polypropylene, but is used as a nucleating agent in HDPE. However, benzoic acid salts and *bis*-benzylidene sorbitol are effective nucleating agents in PP. *p*-Aminobenzoic acid is used as a nucleating agent in polybutene.

## ODORANTS

While scents have been used in some polymers for years to mask undesirable odors, other fragrances are now being used to provide fruit-scented molded kitchen articles, toys, and bottle caps. Some deodorants for trash bags are used in conjunction with biocides.

## PLASTICIZERS

Natural polymers, such as cellulose and proteins, have been flexibilized by water, oils, waxes, and balsams for centuries. The Plastics Age began in 1868, when Hyatt and Parkes flexibilized cellulose trinitrate by the addition of nonvolatile materials such as camphor. Even so, the significance of this advance, which resulted in the production of Celluloid, was overlooked for at least a half century.

Triphenyl phosphate was proposed as a substitute for camphor in 1912, and esters of phthalic acid were produced in the 1920s. The development of the vinyl plastics industry was delayed until Semon and co-workers plasticized PVC with dibutyl phthalate and tricresyl phosphate in 1933. This intractable polymer was also flexibilized in the early 1930s by the copolymerization of vinyl chloride and vinyl acetate (Vinylite). Despite the importance of this internal plasticization, the emphasis in this chapter will be on external plasticizers.

A relatively small amount of external plasticizers is used to improve the flexibility of cellulose esters and acrylics, but the major use is for the plasticization of PVC. Small amounts of plasticizers will actually increase the modulus (stiffness) of PVC. This effect is called antiplasticization.

According to the viscosity theory developed by Leitech in 1943, plasticizers function by modifying the rheological properties of polymers; thus, those having low temperature coefficients of viscosity have mechanical and electrical properties that are less sensitive to temperature. The relationship of kinematic viscosity (v) to Kelvin temperature is shown by the following equation in which the constant C is related to polymer structure and m is the viscosity temperature coefficient:

$$\text{Log } v = 10CT_m$$

According to Flory and Huggins, the polymer plasticizer interaction is dependent on the free energy ($\Delta G$) of the composite. Thus, phase separation occurs if the value of $\Delta G$ is greater than 0. A single phase is usually present when the difference in Hildebrand solubility parameters ($\Delta \delta$) of the plasticizer and the polymer are less than 1.8 Hildebrand units (H).

According to the lubricity theory, the plasticizer reduces the inter-molecular friction between the polymer chains and allows the macro-molecules to slide by each other. Doolittle proposed a gel theory in which the aggregation-disaggregation equilibrium present in the unplasticized polymer is modified by a solvation-desolvation equilibrium. Another theory, known as the free volume theory, assumes that movement of polymer chains is dependent on the movement of chain segments into the free space between the molecules. The free volume is increased by the addition of plasticizers, and thus the ease of segment mobility is increased and the glass transition temperature $(T_g)$ is reduced.

The addition of a plasticizer also lowers the hardness, modulus, and tensile strength of a polymer. Thus, plasticizer efficiency may be esti-mated by comparing the amount of additive required to achieve a speci-fied tensile modulus at room temperature. According to Boyer, the glass transition temperature of the plasticized system $(T_g)$ is equal to that of the unplasticized polymer $(T_o)$ less a constant $(\alpha)$ times the concentration of plasticizer $W_1/M_1$ i.e., weight percent:

$$T_g = T_o \alpha W_1/M_1$$

Kelly and Bueche have used a modified Williams, Landel, and Ferry (WFL) equation to show the relationship of the glass transition tempera-ture of the composite $(T_g)_{PL}$ to the $T_g$ of the polymer $(T_g)$ and the plas-ticizer $(T'_g)$ and the volume fraction of the polymer (C). The value of the constant in the equation is $4.8 \times 10^{-4}$.

$$(T_g)_{PL} = \frac{4.8 \times 10^{-4}\, C\, T_g + \alpha(1 - C)\, T_g{'}}{4.8 \times 10^{-4}\, C + \alpha(1 - C)}$$

Unplasticized PVC is about 15% crystalline, and this crystallinity raises the softening point so that the processing temperature approaches the decomposition temperature of PVC. Small amounts of plasticizer will actually increase the modulus of PVC, but larger amounts (35 to 50 phr; see Chapter 3 for a definition of phr) increase the flexibility of this intractable plastic.

As shown in Table 2.9, phthalates account for almost 70% of all plas-ticizers used. The epoxidized esters, which are also heat stabilizers, account for almost 10% of the total volume of plasticizers used.

| Table 2.9  Annual U.S. Use of Plasticizers (1988) | |
|---|---|
| **Plasticizer** | **Consumption, tons** |
| Diisodecyl phthalate | 121,000 |
| Dioctyl phthalate | 261,000 |
| Other phthalates | 78,000 |
| Epoxidized esters | 57,000 |
| Adipates | 28,000 |
| Polyesters | 21,000 |
| Trimellitates | 13,000 |
| Azelates | 5,000 |
| **TOTAL** | **684,000** |

The phthalate plasticizers include di(2-ethylhexyl) phthalate (DOP), diisooctyl phthalate (DIOP), diisononyl phthalate (DINP), and diisodecyl phthalate (DIDP). The volatility of these phthalates decreases as the size of the alcohol group increases. The adipates, azelates, and sebacates provide lower temperature flexibility than the phthalates. Linear polyesters are more permanent than the liquid plasticizers.

Phthalates are used as plasticizers for cellulosics, polyvinyl acetate, and polyurethanes. Sulfonamides are used as plasticizers for nylon and acrylics. Phosphates provide flame resistance, but these plasticizers may be toxic.

Dehydrated castor oil has been proposed for use as a plasticizer for acrylonitrile-butadiene elastomers (NBR) in place of phthalates. The substitution of medium-weight glutarate polyesters for adipic polyesters as plasticizers has been suggested.

Dioctyl isophthalate provides mar resistance and diocyl terephthalate provides permanence in plasticized PVC. Chlorinated polyethylene is used in Europe and ethylene-vinyl acetate the U.S. to increase the flexibility of PVC, but these additives are usually classified as impact modifiers. The annual worldwide production capacity for plasticizers is shown in Table 2.10.

| Table 2.10 Annual Worldwide Production Capacity for Plasticizers | |
| --- | --- |
| **Region** | **Capacity, 1000 tons** |
| European Community | 1820 |
| North America | 1150 |
| Central America | 5 |
| South America | 130 |
| Asia | 715 |
| Australia | 30 |
| Africa | 50 |
| Eastern Europe | 300+ |
| **TOTAL** | **4200+** |

## PROCESSING AIDS

Plasticized PVC was used for more than two decades before Jennings extruded a mixture of rigid PVC and styrene-acrylonitrile (SAN) in 1953. Comparable results were obtained when methyl methacrylate was used to replace acrylonitrile in the additive. The dominant processing aids for PVC today are high molecular weight copolymers of alkyl acrylates and methyl methacrylate. It is of interest to note that the glass transition temperature of the rigid PVC containing these additives is higher than that of PVC. Copolymers of methyl methacrylate, methyl acrylate, and styrene are also used as processing aids.

It is believed that these additives promote fusion and alter the rheology of PVC formulations (compounds). In the initial processing stage, PVC behaves like a polyolefin and heat transfer and fusion are delayed because of reduced adhesion to the metal processing surfaces. Processing aids promote the formation of a clean, smooth, homogeneous melt. The mix also has a higher melt elasticity, which results in a desirable higher die entry pressure, as well as a higher melt strength, which permits drawing of films without fracture and better control of die swell. About 1000 tons of processing aids are used annually in the U.S.

## SLIP AGENTS

Slip agents are internal lubricants that lower the coefficient of friction and prevent plastics from sticking to metal surfaces during processing. The use of slip agents is dependent on their low compatibility, which results in migration of the additives to the surface of polyolefin films. These additives are usually fatty acid esters or amides. Euramide and oleamide are the most effective commercial slip agents for LDPE. They are used in concentrations of 1 to 3 phr primarily in PVC and polyolefin film to prevent the plastic film from sticking to itself.

## SMOKE SUPPRESSANTS

Smoke is associated with the products of combustion and the concentration of particulate matter. This condition is aggravated by some flame retardants, such as halogen compounds. However, mixtures of antimony oxide ($Sb_2O_3$) and zinc borate ($2ZnO \cdot 3B_2O_3 \cdot 3.5H_2O$) are excellent smoke suppressants.

Sodium aluminum hydroxy carbonate ($NaAl(OH)_2CO_3$, Dawsonite) is also an excellent smoke depressant, but is not used since it is alleged to be a carcinogen. However, alumina trihydrate and magnesium hydroxide ($Mg(OH)_2$) are effective flame retardants and smoke depressants when used in high concentrations (35%).

In contrast, other smoke depressants are effective in the following concentrations: zinc oxide, 15%; iron dicyclopentadienyl ferrocene, 2%; and molybdenum oxide, 10%. The high cost of the latter system may be reduced by blending with equal parts of cupric oxide. The most widely accepted test is ASTM E662, in which the polymers are exposed to radiation from a heater and the smoke density is measured by a vertical light beam.

## ULTRAVIOLET LIGHT STABILIZERS

That biopolymers are degraded in sunlight has been recognized for centuries, but because the reactions were not understood, the only recourse was to keep these materials in the shade whenever possible. In the

1950s, Bateman and Norris showed that peroxides were formed so rapidly during photooxidation that antioxidants were not effective in suppressing this degradation. Hence, when it was essential to expose polymers to wavelengths in the ultraviolet spectrum (290 to 400 nm), it was necessary to add ultraviolet absorbers which were more powerful than the carbonyl group (C=O) formed from the hydroperoxy groups produced by exposure to sunlight or fluorescent lighting.

A review of bond energies will show that 84 kcal/mol is sufficient to cleave carbon-carbon bonds; this energy corresponds to 340 nm. The energy is lower for cleavage of the carbon-carbon bonds in ethylbenzene (63 kcal/mol) and in aliphatic ketones such as acetone ($H_3CCOCH_3$) (72 kcal/mole. The pioneer UV stabilizer was phenyl salicylate (salol, $C_6H_4OHCOOC_6H_5$). This additive was first used in suntan lotions.

It is now known that derivatives of esters of salicylic acid, benzoin acid, and phthalic acid with phenol and resorcinol undergo a photo-Fries rearrangement to produce 2-hydroxybenzophenone (HBP), which is an effective UV stabilizer that absorbs in the 260 nm range. 2-hydroxy-benzophenone and its derivatives form chelates via intramolecular hydrogen bonding. These chelates absorb UV energy and form a quinoid structure which reverses back to the original structure by releasing thermal energy. The chelate and quinoid structures are shown in the following equation:

4-alkoxy derivatives of HBP shift the absorption toward the lower end of the UV range (285 nm) and increase the absorptivity over the critical 290 to 400 nm range. The efficiency of HBP absorbers is enhanced as the strength of the hydrogen bond increases and is weakened in highly polar media.

Hydroxyphenylbenzotriazoles, shown in the following structural formula where X is H or Cl, have higher molar extinction coefficients and absorption in the 400 nm range. The mechanism for these UV additives is similar to that described for HBP.

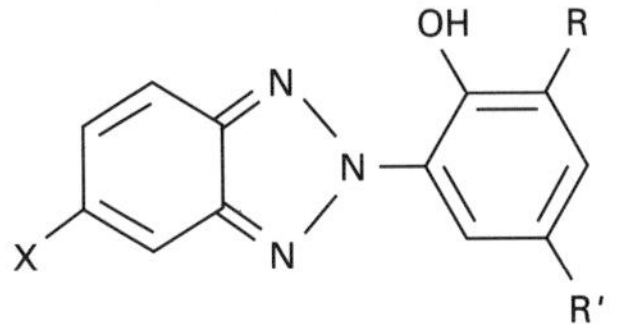

Derivatives of cinnamic acid ($C_6H_5-CH=CHCOOH$) have zero absorptivity in the visible region of the electromagnetic spectrum and hence do not affect optical clarity. However, they are effective only in the 310 to 320 nm range and have limited use as UV stabilizers.

Another group of UV stabilizers interacts with photoexcited macromolecules and dissipates the excess energy in the infrared range. The formula for a nickel chelate used as a commercial excited state quencher is shown below:

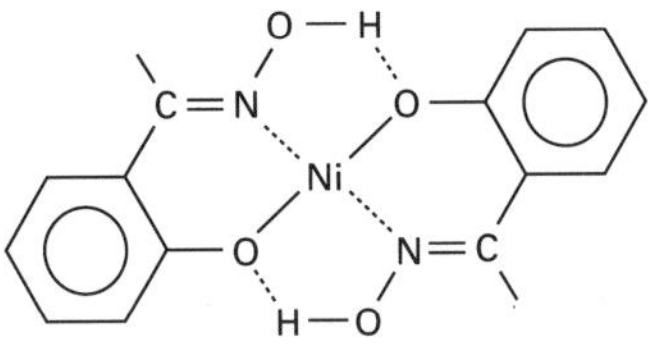

Polymers may also be light stabilized by the addition of hydroperoxide decomposers, such as dialkyldithiocarbamates, and by free radical scavengers. A recent development in this field is illustrated by hindered amine stabilizers (HALS), such as *bis*-(2,2,6,6-tetramethyl-4-piperidyl) sebacate, which cannot be classified as a UV absorber or excited-state quencher since it does not absorb radiation at wavelengths greater than 250 nm.

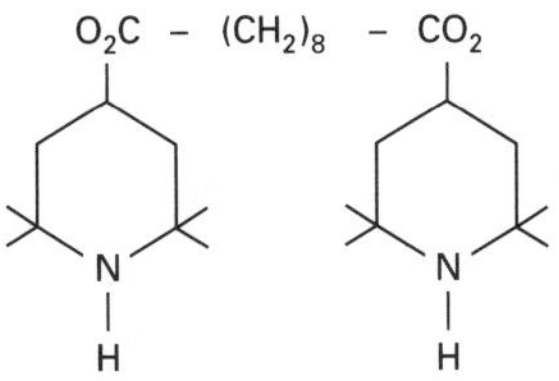

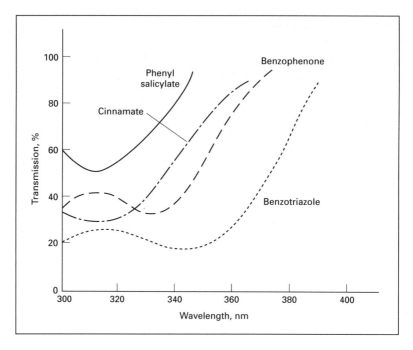

*Fig. 2.1 Percent transmission of major types of UV stabilizers*

It is believed that HALS are converted to nitroxyl radicals (>N–O·), which couple with free radicals. HALS and other UV stabilizers are effective in low concentrations (0.5%).

Zinc oxide and titanium dioxide and mixtures of zinc oxide and dialkyl thiocarbamate are effective UV stabilizers. Of course, carbon black is also effective in applications in which a black color is acceptable.

The UV transmission of the major types of stabilizers is shown in Fig. 2.1. More than 2000 tons of UV stabilizers are used annually by the American polymer industry.

## THIXOTROPES AND
## VISCOSITY CONTROL ADDITIVES

The flow and sagging of polyester prepolymers can be controlled by the addition of thixotropes, which increase the viscosity of static fluids.

Organosilanes and titanates reduce the viscosity of filled polymers, and viscosity depressants, such as ethoxylated fatty acids, can be used to reduce or control the viscosity of PVC plastisols. High concentrations of fillers, coated with coupling agents, can be added to plastisols without a noticeable increase in viscosity.

The American plastics industry compounded 5.8 million tons of composite plastics in 1989. It is anticipated that this volume will increase to 8.7 million tons by 1994. The growth rate pacesetters will be conductive plastics and plastic blends. As shown by the data in the following section, tolling custom and color compounding will account for more than 90% of the U.S. compounding business.

## PRODUCTION AND COST DATA FOR ADDITIVES

Frost and Sullivan predict a 900 million pound plastics additives market with an estimated value of $1.2 billion in 1994. This predicted volume in millions of pounds and price in millions of dollars is:

| Additive | Volume | Price |
|----------|--------|-------|
| Flame retardants | 647 | $473 |
| Colorants | 50 | $221 |
| Heat stabilizers | 113 | $186 |
| Antioxidants | 62 | $150 |
| Light stabilizers | 11 | $94 |
| Blowing agents | 15 | $34 |

According to M.C. McMurrer, the 1986 U.S. consumption of additives, in thousand tons, was as follows:

| Plasticizers | 850 |
| --- | --- |
| Colorants | 173 |
| Blowing agents | 107 |
| Flame retardants | 100 |
| PVC impact modifiers | 57 |
| Heat stabilizers | 49 |
| Lubricants | 45 |
| Carbon black | 35 |
| Organic peroxides | 21 |
| Antioxidants | 20 |
| UV stabilizers | 4 |
| Urethane catalysts | 3 |
| **TOTAL** | **1464** |

## FILLERS

Some pigments, flame retardants, and other additives discussed in this chapter also behave as fillers in polymers. These and other fillers will be discussed in Chapter 3.

## REFERENCES

- K.E. Atkins *et al.*, *Plast. Comp.*, Vol. 11 (No. 4), 1988, p. 35
- K.E. Atkins *et al.*, *Plast. Comp.*, Vol. 11 (No. 5), 1988, p. 35
- D. Balmer, *Plast. Comp.*, Vol. 13 (No. 2), 1990, p. 38
- T.J. Blong and D. Duchesne, *Plast. Comp.*, Vol. 13 (No. 1), 1990, p. 50
- L. Bohlin, G. Malhammar, and H.E. Stromvall, *Plast. Comp.*, Vol. 13 (No. 1), 1990, p. 31
- L.A. Canoza, S.C. Staffore, and A.D. Ulrich, *PMAD (SPE)*, Vol. 5 (No. 6), 1987, p. 7
- D. Davidson and P. Stewart, *PMAD (SPE)*, Vol. 3 (No. 1), 1985, p. 8
- F.J. Fletcher, T.C. Dean, and A. Docherty, *PMAD (SPE)*, Vol. 1, 1994, p. 6
- R. Gachter and H. Muller, *Plastics Additives Handbook*, Hanser Publishers, New York, 1987

- L. Hanlan, *Plast. Comp.*, Vol. 13 (No. 2), 1990, p. 47
- P.R. Hornsby and C.L. Watson, *PMAD (SPE)*, Vol. 7 (No. 3), 1990, p. 7
- B. Johnson and J. Kunde, *PMAD (SPE)*, Vol. 6 (No. 1), 1988, p. 7
- D.M. Kalyon and M. Khemis, *PMAD (SPE)*, Vol. 1 (No. 1), 1985, p. 5
- J.I. Kroschwitz, Ed., *Encyclopedia of Polymer Science and Engineering*, Vol. 2, Wiley-Interscience, New York, 1985
- R.A. Lindner, *Plast. Comp.*, Vol. 13 (No. 2), 1990, p. 58
- N.B. Martin, H. Tomlinson, and T. Brewer, *PMAD (SPE)*, Vol. 1 (No. 1), 1985, p. 5
- L. Mascia, *The Role of Additives in Plastics*, Edward Arnold, Ltd., London, 1974
- M.C. McMurrer, *Plast. Comp.*, Vol. 11 (No. 4), 1988, p. 8
- S.J. Monte and G. Sugarman, *Plast. Comp.*, Vol. 12 (No. 7), 1989, p. 59
- M. Morton, Ed., *Introduction to Rubber Technology*, Reinhold, New York, 1959
- B. Mulholland, *PMAD (SPE)*, Vol. 6 (No. 2), 1988, p. 8
- T.C. Patton, *Pigment Handbook*, Vol. III, Wiley-Interscience, New York, 1973
- T.C. Pedersen, *PMAD (SPE)*, Vol. 1 (No. 1), 1984, p. 7
- J.A. Radosta, *PMAD (SPE)*, Vol. 1 (No. 1), 1984, p. 6
- T.D. Ritchie, Ed., *Plasticizers, Stabilizers, and Fillers*, The Butterworth Group, London, 1972
- D. Sekutowski, *PMAD (SPE)*, Vol. 5 (No. 2), 1987, p. 5
- R.B. Seymour, Ed., *Additives for Plastics*, Academic Press, New York, 1978
- K.K. Shen, *Plast. Comp.*, Vol. 11 (No. 7), 1988, p. 26
- R.G. Webber, Ed., *Coloring of Plastics*, John Wiley & Sons, New York, 1979
- A. Whelan and J.L. Crafts, *Developments in Plastics Technology*, Vol. 2, Elsevier Applied Science, London, 1985
- G.G. Winspear, *The Vanderbilt Rubber Handbook*, R.T. Vanderbilt Company, New York, 1968
- A.S. Wood, *Modern Plastics*, Vol. 67 (No. 5), 1990, p. 40
- R.T. Woodhans and I. Chun, *PMAD (SPE)*, Vol. 1, 1984, p. 5

# CHAPTER 3

# Fillers

## INTRODUCTION

The use of fillers has brought about a major change in the plastics industry. Fillers have been an integral part of thermosets for almost a century, but the use of filled thermoplastics is relatively new. According to McMurrer, 2.5 million tons of mineral-, synthetic-, and organic-based fillers were used in the U.S. in 1988; it is anticipated that this volume will be about 3.5 million tons in 1993. Some pigments and some flame retardants (see Chapter 2) may also be considered as fillers when used in relatively large concentrations. These mixtures of polymers and fillers may be classified as plastic composites.

Fillers have been used inadvertently in natural rubber, paints, and Celluloid for over a century. However, the first-large scale use was for the reinforcement of phenolic resins by attrition ground wood (wood flour) by Baekeland in the early 1900s. As is the case with Bakelite, fillers usually increase hardness and mechanical properties and reduce shrinkage, but increase the specific gravity of the composite.

Since nonresinous particles obstruct resin flow, filler particle morphology must be considered in the design of a plastic composite. The resistance to flow is greater for composites with acicular (needlelike) particles than for those with more regular shapes, such as spheres. The shape of the particle is related to the ratio of its length to diameter, or aspect ratio. The term "loading" is used to describe the concentration of fillers present in a composite. However, terms such as parts of filler per 100 parts of resin (phr), weight percent, and volume percent (pvc) are also used.

Some fillers, such as carbon black and surface-treated fillers, are actually bonded to the macromolecular chains in the matrix. Others may be considered as nonreactive fillers, but they do immobilize the polymer chains as attested to by an increase in glass transition temperature ($T_g$). However, the nonreactive fillers have less effect than more reactive fillers on modulus and heat deflection temperature of the composites.

## THEORIES OF REINFORCEMENTS

It is generally believed that external stresses on a plastics composite are transferred from the continuous phase (resin) to the discontinuous phase (filler). Thus, the ultimate properties of the composite are dependent on the extent of bonding between the two phases, the surface area of the filler, and the extent of packing of the filler particles.

Suppliers of fillers usually provide information on surface area which may be obtained by the Brunauer-Emmet-Tellor (BET) equation, which is based on the isothermal adsorption of nitrogen or by methylene blue. The art of packing, in which the voids formed by coarse particles are filled by finer particles, which has been used for over a century in making concrete, has also been applied to polymer composites.

In the BET equation, the partial pressure of the adsorbate (F) is a function of the volume of nitrogen adsorbed (V), the volume of nitrogen corresponding to a monolayer of gas ($V_m$), and a constant (C) which is a function of temperature and the difference between the heats of liquefication and monolayer adsorption:

$$\frac{F}{V(1-f)} = \frac{1}{V_m C} + \frac{(C-1)F}{V_m C}$$

Information on the average number of adsorbed and nonadsorbed polymer segments, the average segment density, and the mean thickness of the adsorbed layers of polymer on the filler surface may be calculated by use of statistical mechanics. These and Monte Carlo calculations show that the thickness of the adsorbed layer is related to the adsorption energy and that the average number of adsorbed and nonadsorbed segments are not affected by chain length at moderate energy adsorption below C values.

The packing properties ($P_f$) of fillers can be estimated from the volume of oil adsorbed (ASTM D281). However, $P_f$ values will be affected by polarity, viscosity, and molecular entanglement. Nevertheless, the ratio of fractional volume of filler in the composite $V_f/P_f$ is the most significant value in predicting properties of the composite. The tensile strength, flexural strength, and impact strength of the composite are all related to $V_f/P_f$ values.

Composite science has trailed the application of composites, but some properties can be estimated from the $V_f/P_f$ values and by use of the Einstein equation, in which the ratio of the viscosity of a mixture of suspended spheres in a liquid ($\eta_o$) to that of the compounded liquid ($\eta$) is related to the Einstein constant (C), which is 2.5 for spheres whose aspect ratio, by definition, is equal to 1.0.

The Einstein equation has been modified by Guth and Gold in the EGG equation, which may be extended to show the ratio of moduli (E) which are related to viscosities:

$$\frac{E}{E_o} = \frac{\eta}{\eta_o} = 1 + 2.5C + 14.1C^2$$

## SPECIFIC FILLERS

Fillers have been classified in accordance with source, function, and importance. However, for reasons of simplicity, they will be discussed in alphabetical order in this chapter. Several examples of applications for mineral-filled plastics are shown in Fig. 3.1.

### Aluminum Flakes

Aluminum flakes, with an aspect ratio of about 40, have been added to polymers in small amounts (15 vol%) for electrostatic dissipation, in moderate amounts (25 vol%) for shielding from electromagnetic interference (EMI) and radio frequency interference (RFI), and in larger amounts for thermal conductivity (40 vol%). Plastics with a 20% loading of aluminum flakes have a heat transfer effectiveness of 20% of that of pure aluminum and thus are effective for EMI shielding.

### Alumina (Aluminum Oxide)

Corundum ($Al_2O_3$), boehmite ($Al_2O_3 \cdot H_2O$), and gibbsite ($Al(OH)_3$) are naturally occurring minerals. Bauxite, which has a high content of gibbsite, is the principal source of aluminum. This mineral is heated with sodium hydroxide to obtain soluble sodium aluminate. The insoluble residue, which is called "red mud," contains sodium aluminum silicate. The "red mud" is used in the Republic of China (Taiwan) and other Asian countries and as a filler for polyvinyl chloride (PVC). Most of the soluble

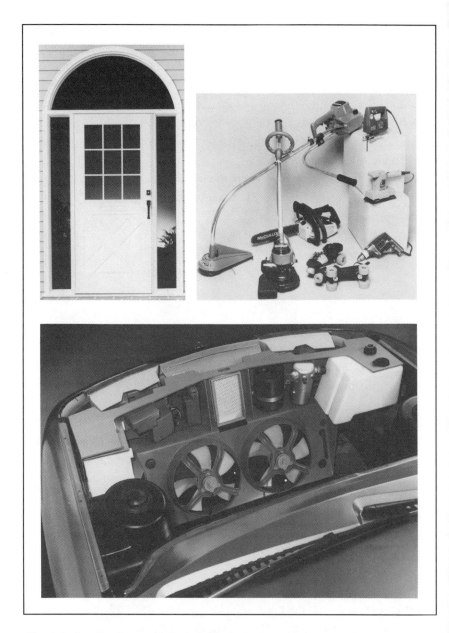

Fig. 3.1 Applications of mineral-filled plastics. (Clockwise from upper left):
Plastic storm door (G.E. Plastics); plastic tools and skates (Allied-Signal
Corp.); plastic composite automotive front-end module (G.E. Plastics)

sodium aluminate is used to produce aluminum metal by electrolysis, but some is used to make alumina filler and DuPont's fiber FP. Alumina has a Mohs' hardness of 9 and a specific gravity of 4.0.

The most widely used alumina filler is gibbsite ($Al(OH)_3$), which is usually called alumina trihydrate (ATH). As stated in Chapter 2, this nontoxic white filler, which has a Mohs' hardness value of 3 and a specific gravity of 2.4, is widely used (in high concentrations) as a flame retardant. Since it decomposes endothermically to produce ($Al_2O_3$) and water (34.6%), composites containing this additive must be processed at temperatures below 220 °C.

Because the index of refraction of ATH (1.57) is similar to that of many polymers, ATH composites are often translucent. This amphoteric filler forms weak bonds with many polymers; the strength of these bonds and the filler loading can be increased by surface treatment with coupling agents. Almost 100,000 tons of ATH are used annually by the American polymer industry.

## Aluminum Silicate

As stated in the previous section, "red mud," which contains aluminum silicate, is used as a filler for PVC in Taiwan and other Asian countries. Synthetic sodium aluminum silicate (SSAS) is produced by the reaction of aluminum sulfate with sodium silicate. When the oil adsorption values exceed 200 $cm^3/1000$, these silicates are called "very high structure" (VHS). SSAS is used both as a filler and a pigment or color extender.

## Antimony Oxide

Antimony oxide ($Sb_2O_3$) is mined in Algeria and Mexico. However, antimony oxide, which is used primarily as a flame retardant by the plastics industry, is a white powder obtained from antimony sulfide (stibnite) or antimony oxide ores. About 15,000 tons of this additive are used annually by the American plastics industry.

## Asbestos

There are five types of amphibole and one type of serpentine or chrysotile asbestos. The latter is the most widely used type, but crocidolite has been used, to a small extent, as a filler. Asbestos was used by Baekeland to provide dimensional stability, durability, improved impact strength, and heat resistance to phenolic resins. It has also been widely used in PVC floor tile.

Because of the danger of asbestosis, bronchogenic cancer, mesothelioma, and gastrointestinal cancer associated with exposure to asbestos, the Occupational Safety and Health Administration (OSHA) has placed severe limitations on exposure to asbestos in the air. Hence, despite the significant increase in modulus, flexural strength, and heat deflection temperature of asbestos-filled plastics, the use of asbestos filler has decreased.

One advantage of asbestos over other high-aspect fillers is flexibility. Thus, while fiber bundles may be subdivided during processing, there is very little decrease in fiber length. Over 235,000 tons of asbestos are used annually by the American plastics industry.

## Barium Sulfate (Barite)

Barium sulfate ($BaSO_4$) is a naturally occurring mineral mined in the western U.S. This additive has been used as a white pigment in paints, in thixotropic oil drilling muds, and as an X-ray opaque filler in plastics. This filler, which has a Mohs' hardness of 3 and a specific gravity of 4.3, is used as a filler in PVC extrudates which are attached to surgical sponges so that their presence may be detected by X-rays in surgical operations. About 3000 tons of barium sulfate are used annually by the American plastics industry.

## Beryllium Oxide

Beryllium oxide, which occurs as the mineral beryl, is usually produced commercially by thermal decomposition of beryllium hydroxide or nitrate. This filler, which has a Mohs' hardness of 7 and a specific gravity of 2.9, improves the heat conductivity of polymeric composites. This element was originally called gluconium because its salts have a sweet taste; however, beryl dust can cause berylliosis, which is often fatal after a delayed exposure time.

## Calcium Carbonate (Limestone)

Calcium carbonate occurs throughout the world as sedimentary rocks of marine origin. It is the most widely used and least expensive filler in the polymer industry. Calcium carbonate occurs as chalk (calcite), limestone, and marble. Precipitated calcium carbonate filler is also produced by the addition of carbon dioxide ($CO_2$) to a solution of calcium chloride ($CaCl_2$). Dolomite, which is also used as an inexpensive filler, is a double carbonate of magnesium and calcium ($CaMg(CO_3)_2$).

Calcium carbonate has a Mohs' hardness of 3, a specific gravity of 2.7, and a refractive index of 1.48 to 1.66. Calcium carbonate fillers are widely used in phenolic resins. Calcium carbonate is also used as a filler in PVC plastisols, PVC organosols, rigid PVC, and PVC floor tile.

The bonding of plastics to calcium carbonate fillers can be improved by surface treatment with stearic acid, alkyl sulfonic acid, or titanate or zirconate coupling agents. About 680,000 tons of calcium carbonate are used annually by the American plastics industry.

## Carbon Black

Carbon black has been produced for over 2000 years. It was used for centuries as a pigment in India ink, and Charles Goodyear patented its use as a pigment in rubber in the 1840s. However, its use as a reinforcing filler for rubber was not recognized until 1904, when S.C. Mote used carbon black and cord reinforcements for pneumatic tires in Silvertown, England.

The use of carbon black as a reinforcing filler in the U.S. was the result of an analysis of the Silvertown tires imported from England, but no attempt was made to patent this filler. Hence, all U.S. tire firms adopted this innovation after 1910. W.B. Wiegand called carbon black the "king of pigments." He proposed a mechanism for reinforcement and demonstrated that the extent of reinforcement varied with the particle size of the filler.

The effectiveness of carbon black as a filler is demonstrated by the average mileage of service of pneumatic tires before and after the incorporation of carbon black in tires in 1910. The average tire life in 1900 was 500 miles (805 km); in 1912 it was 3000 miles (4830 km); and in 1978 it was 38,000 miles (61,180 km).

Impingement (channel) black is produced by burning natural gas, with a limited supply of air, and allowing the carbon particles to be deposited on a cool metal (channel) surface. While this process was acceptable to the ancient Chinese and to some primitive American producers, it was not acceptable to those living within 10 km of the channel black plants nor to the Environmental Protection Agency (EPA). Thus, channel black

| Table 3.1  Leading U.S. Producers of Carbon Black | |
| --- | --- |
| **Producer** | **Production, 1000 tons** |
| Cabot | 327 |
| Columbia | 276 |
| Ashland | 254 |
| Phillips | 195 |
| J.M. Huber | 191 |
| Witco | 125 |
| Sid Richardson | 91 |
| Gulf | 9 |
| Union Carbide | 3.5 |

has been displaced almost entirely by furnace black, which is produced by burning oil with a smoky flame in a closed system.

In addition to functioning as a reinforcing filler in pneumatic tires, carbon black also functions as a UV stabilizer, antistat, colorant, thermal antioxidant, and conductive filler in plastics. The use of graphite flakes as a lubricant in styrene-butadiene elastomers (SBR) has also been proposed. The mean particle size of carbon black is as follows: furnace black (13 to 70 μm), channel black (10 to 30 μm), thermal black (50 to 500 μm), lamp black (50 to 100 μm), and acetylene black (350 to 500 μm). The so-called energy-intensive furnace process black made by a gas furnace process has been replaced, to a large extent, by oil furnace black, which is the preferred filler for SBR. The thermal process is a cyclic thermal decomposition process which, because of the high cost of natural gas, has been displaced by oil furnace black.

The annual production of carbon black in the U.S., in thousands of tons, is as follows: furnace black, 1035; thermal black, 18; acetylene black, 7; lamp black and replacement lamp black, 3. The U.S. capacity for furnace black is 1,580,000 tons, and the worldwide capacity for this carbon black is 5,603,000 tons. The leading U.S. producers of carbon black are shown in Table 3.1.

Lampblack, which is produced by collecting the soot from burning petroleum on a cool surface, has a large particle size and "high structure" and is used, to a small extent, by the coatings industry. The extent of aggregate formation (graping effect) of carbon black as reticulate chains is called "structure" in the rubber industry. "Structure," which affects the physical properties of the polymer composite, is also related to "bound rubber." Bound rubber, i.e., the rubber that is adsorbed from solution on the carbon black surface, is related to the total interfacial area of filler and polymer. The reinforcing factor (RF) of a carbon black may also be estimated by measuring the change in flow of a solution of polymer and filler in a capillary rheometer.

Hollow carbon spheres (microspheres) are produced by heating a slurry of carbon black and resin in a vacuum. These spheres are used in syntactic foams. Ground petroleum coke is used as a filler in polymer cements. Carbon filaments are produced by the pyrolysis of acrylic fibers or pitch.

## Clay

Clay is a hydrous aluminum silicate in which alkalies or alkaline earths are also present. Clay may be amorphous (allophane), crystalline (kaolinite, halloycite), montmorillonite, vermiculite, mixed-layer type (chlorite) or chain structured (attapulgate). These minerals, which are widely distributed geographically throughout the world, are the natural decomposition products of feldspar, and include kaolins (kaolinite, dickite, nacrite) with the formulas $Al_2O_3 \cdot 2SiO_2 \cdot 2H_2O$ and $Al_2O_3 \cdot 2SiO_2 \cdot 4H_2O$. Other clays include serpentines, which contain magnesium $(Mg_3SiO_5(OH)_4)$; smectites (montmorillonites), which carry a lattice charge; and illites, which are constituents of argillaceous sediments, glauconate micacean clay, attapulgite, and sepiolite.

Delaminated, water-washed, and calcined clays are available in a variety of particle sizes. When added to polyester prepolymers, clay acts as a heat sink, which results in a reduction in the peak exothermic temperature and decreases surface crazing. The polyester composites containing delaminated clay have higher tensile strengths and moduli than composites containing standard kaolin.

Kaolin (china clay) has a Mohs' hardness of 2, an index of refraction of 1.56, and a specific gravity of 2.6. Kaolin is one of the most widely used fillers in polymers. It is used in PVC, nylon, rubber, polyolefins, polyesters, and polyurethanes. Fully calcined clays are white, while the partially calcined clays are tan in color. Calcined clays impart good electrical properties to composites and are used in the wire and cable industry. All grades of clay may be treated with surface modifying agents to obtain improved dispersibility and water resistance when used in polymer composites.

Kaolin (hydrous aluminosilicate) is widely used as a filler in plastics, paper, and elastomers. The naturally occurring platelets may be delaminated by cleavage of the kaolin stacks, and this filler may be dehydroxylated by calcination. Clay is the least energy intensive of the mineral fillers and is readily dispersed in resins, but is rarely used as the sole filler. The concept of ionization potential of clay has been used to predict chemical bonding of this filler to the resin.

A synthetic sodium aluminum silicate (SSAS) is produced by the reaction of sodium silicate and aluminum sulfate. Feldspar and nepheline are also aluminum silicates, obtained by crushing granite and syenite, respectively. Because of a high index of refraction of 1.53, these heat-stable fillers may produce translucent composites. More than 80,000 tons of clay are used annually by the American polymer industry.

## Glass

In addition to fiberglass, other forms of glass, such as glass ribbons, milled glass fibers, glass flakes, and both solid and hollow glass beads, have been used in polymeric composites. Glass ribbon has found limited use in polymer composites. When added to polyurethanes, glass flakes provide a hard scratch- and corrosion-resistant surface.

Fibers produced by milling chopped glass fibers into shorter lengths have the full length to diameter of the filler available for reinforcement. Milled fiber and glass spheres are isotropic, and therefore stresses around these particles are evenly distributed. When used as a replacement for fibrous glass, solid glass spheres improve flow and reduce the cost of bulk molding compounds (BMC). The dynamic and transient rheological properties of glass-filled melts are also improved by the addition of glass spheres.

Glass microspheres improve the strength, stress distribution, and surface finish of polymer composites. Because sharp edges are absent, there is minimal wear on processing equipment, despite the relatively high Mohs' hardness of 6. However, glass sphere-filled parts, such as acrylonitrile-butadiene-styrene (ABS) watering cups used in egg factories, have inherent abrasion resistance and last 20 times longer than unfilled ABS cups.

Hollow glass beads or bubbles are now being used at high loadings in reaction injection molding (RIM). These hollow glass beads, available in a wide variety of sizes, improve impact resistance and reduce the weight of composites. Polymers such as polystyrene filled with hollow glass spheres are called syntactic foams. These spheres are also used with thermosets, such as epoxy and polyester resins.

The specific gravity of glass microspheres ranges from 0.12 to 0.7; the average diameter is 65 to 75 μm, with a wall thickness of 1 to 2 μm. These spheres also act as nucleating agents in syntactic foam. A "short game" golf ball, which travels one-half the distance of a conventional golf ball, is made from hollow glass microsphere-filled ionomers.

Since glass microspheres are hygroscopic, they are often coated with silanes, silicones, or other coupling agents. These spheres are available in sizes ranging from 325 to 20 mesh. The surface area (SA) (in $cm^2/kg$), which is dependent on size gradations, is inversely related to diameter (D) (in μm), as shown by the following equation:

$$SA = \frac{500 \times 10^5}{D}$$

About 16,000 tons of glass spheres are produced annually in the U.S. from molten A glass.

## Iron Oxide

Both natural iron oxides (ochers, umbers, and siennas) and synthetic iron oxides are used as pigment and fillers. Magnetic fillers are produced by heating iron oxide ($Fe_2O_3$) with oxides or carbonates of barium, lead, manganese, strontium, or zinc. The ferrites obtained from manganese, zinc, or blends of the two are termed soft ferrites, while barium ferrite and strontium ferrite are called hard ferrites. Flexible magnets are obtained when these ferrites are added to flexible polymers.

## Lead

Powdered lead is added to flexible polymers, such as polyolefins, silicones, polyurethanes, and plasticized PVC to obtain flexible composites that are X-ray opaque.

## Magnesium Oxide (Magnesia)

Magnesium oxide (MgO), which is produced by heating magnesium chloride ($MgCl_2$) with lime ($CaCO_3$), is a white powder with a specific gravity of 3.6. This filler is used to increase the viscosity of unsaturated polyester premixes and increase the modulus and hardness of resins, such as polypropylene.

## Metal Powders

In addition to aluminum and lead fillers, discussed previously in this section, copper, molybdenum, tin, and tungsten metal powders are used in composites to increase abrasion resistance and machinability and to improve conductivity. Both the ultimate stress and strain of these composites are inversely proportional to the particle size. The thermal conductivity is related to the volume fraction of the fillers. The annual production of metal powders in the U.S., in thousands of tons, is as follows: aluminum, 28; copper, 19; iron and steel, 216; molybdenum, 2; tin, 8; and tungsten, 2.

## Mica

Mica is a lamellar (flakelike) filler consisting of naturally occurring muscovite ($K_2Al_4(Al_2Si_6O_{20})(OH)_4$) or phlogopite ($K_2(MgFe)_6Al_2Si_6O_{20}$) $(OH,F)_4$). Mica-filled plastic composites are characterized by high stiffness, high dimensional stability, and good electrical properties. The low hardness of mica (2.5 to 3.0 Mohs) results in less wear on processing equipment than that encountered with composites with harder fillers, such as glass (6 Mohs).

Care must be exercised when processing composites with mica having high aspect ratios (HAR). At high loadings, mica-filled polymeric composites have a modulus comparable to aluminum and also exhibit a metallic sound when dropped.

Mica has been used as a filler in polyurethanes to reduce thermal expansion and in polypropylene to increase flexural modulus and impact strength. Mica has also been used as a filler in engineering resins, polyesters, nylon, and polyurethanes. It has been stated that the full potential of phlogopite mica in thermoplastics will not be realized until a solution is reached on the knit-line problem. The knit-line is the line formed during injection molding of composites when two streams of the molten material melt and cool in the mold. There is a minimal mixing of the two streams on both sides of the knit line.

More than 185,000 tons of mica are used annually in the U.S., but only about 10,000 tons are used by the plastics industry. Most of the mica produced is used in plaster board, cement, and in asphalt roofing compositions. However, the demand for high-density mica-filled plastics is increasing.

## Molybdenum Disulfide (Molybdenite)

Molybdeum disulfide ($MoS_2$) is a naturally occurring black filler with a specific gravity of 4.8, a Mohs' hardness of 1, and a coefficient of friction of 0.04. It improves processibility and increases the physical properties of composites. Its distinct characteristic is its lubricity or slipperiness.

## Silicas

Silica and its previously discussed derivatives, e.g., talc, clay, and other silicates, account for much of the earth's surface. Silica ($SiO_2$) occurs as quartz, tridymite, cristobalite, and polyamorphous forms of these silicas. Microcrystalline silica is smaller in size but similar to crystalline silica. Quartzite, sand, sandstone, novaculite, and tripoli (microcrystalline silica) are microcrystalline minerals. Microamorphous silica is classified as microparticulate silica, macroscopic sheets and fibers, and hydrated amorphous silica. Pyrogenic (fumed silica) and precipitated silica are included in the microparticulate class.

The coarse grains of crystalline silica permit heavy resin loadings and excellent dimensional stability of moldings and potting compounds. Microcrystalline silicas are less abrasive than ordinary silica and are used as fillers for thermosets, silicone rubber, and thermoplastics.

Precipitated silica may be produced *in situ* by the addition of ethylamine or aqueous ammonia to tetraethylorthosilicate. The morphology of the precipitated silica is a function of the concentration of the amine, water, and ethanol used in the precipitation step. The efficiency of precipitated silica as a reinforcing agent is inversely proportional to its size.

Diatomaceous earth is primarily an amorphous silica produced from skeletons of planktons. Fused or pyrogenic silica is produced by the combustion of silicon tetrachloride in the presence of hydrogen and oxygen. Colloidal silica is produced by the displacement of water by ethanol in aqueous solutions of sodium silicate. Quartz, which has a specific gravity of 2.6 and a Mohs' hardness of 7, is the hardest of the common minerals.

Thixotropic agents with thixotropy comparable to fumed silica are now prepared by the acidic precipitation of silica. Precipitated silica has been used to reinforce silicones, epoxies, polyesters, PVC, and ethylene co-polymers and as a coagent for ATH in flame-retardant applications. Poly-methyl methacrylate sinks and tubs are being reinforced by silica.

Precipitated silica is used as a reinforcing filler in natural rubber, SBR, and silicone elastomers. About 50,000 tons of silica are used annually by the American polymer industry.


## Talc

Talc is a naturally occurring hydrated magnesium silicate which in its pure form has the formula ($3MgO \cdot 4SiO_2 \cdot H_2O$). It occurs as a brucite sheet ($Mg(OH)_2$) sandwiched between two identical tetrahedral sheets. The talc plates are held together by Van der Waals' forces, but some ionic bonding between contaminates may be present.

Talc has a Mohs' hardness of 1, a specific gravity of 2.7, an index of refraction of 1.57, hydrophobic surface properties, and a slippery feel. Crystalline talc may be lamellar, fibrous, foliated, or massive.

Because of its plateletlike structure, talc imparts greater stiffness, tensile strength, and creep resistance than particulate fillers. When heated at 380 to 500 °C, talc loses some of its associated water and loses its combined water of crystallization at 800 °C as it dissociates exothermi-cally into enstatite. Talc-filled polypropylene is used widely in automobile and appliance applications. Other talc-filled composites are based on high-density polyethylene (HDPE), polystyrene, and thermoplastic elas-tomers (TPE).

The use of this white filler is expected to grow at an annual rate of 8%. This platy, low-cost filler, which is usually white in color, is used in polyolefins and polystyrene. When this hydrated magnesium silicate filler is used to replace one-half of the fiberglass in nylon 6, the heat-deflection temperature of the composite is increased by 50 °C. The swelling of acrylonitrile-butadiene rubber (NBR) in solvents is reduced by the addition of talc. Since excessive inhalation of talc dust can cause respiratory disease, exposure to talc dust must be minimal.

About 45,000 tons of talc are used annually by the American polymer industry, and this volume will increase in the future.

### Titanium Dioxide
As stated in Chapter 2, titanium dioxide ($TiO_2$) is the most widely used white pigment. The anatase form has a specific gravity of 3.8, a Mohs' hardness of 5.5, and an index of refraction of 2.5. The rutile form has slightly higher values for these properties: specific gravity of 4.3, Mohs' hardness of 6, and refractive index of 2.7. The anatase form is used for delustering nylon film and in white sidewalls of pneumatic tires. The annual worldwide production capacity of titanium dioxide is about 3 million tons.

### Wallastonite
Wallastonite is a naturally occurring acicular (needlelike) calcium meta silicate ($CaSiO_3$) mined in Willsboro, NY. This white reinforcing filler has an aspect ratio of 20/1, a specific gravity of 2.9, a Mohs' hardness of 4.5, and an index of refraction of 1.63. It has a coefficient of expansion of $6.5 \times 10^{-6}$ cm/cm/°C and changes to pseudo-wallastonite at 1200 °C. The principal application of wallastonite is in polyurethanes and polystyrenes.

### Wood Flour
Wood flour is hygroscopic and dark in color but has a low specific gravity. Its use in thermosets is decreasing, but it is being used as a filler with polypropylene (Gorstock). This filler, which is obtained by attrition or buhr mill grinding of debarked wood with low resin content, is used in the U.S. at an annual rate of 113,000 tons.

Nut shell flour is used at an annual rate of 3000 tons. Starch is used as a biodegradable filler in plastics. About 20,000 tons of alpha-cellulose is used annually for the reinforcement of urea and melamine resins. Short cellulosic fibers are used for rubber reinforcement. Jute has also been used for this application, but it has poor adhesion to rubber.

### Zinc Oxide
Zinc oxide (ZnO), which is produced by the oxidation of zinc vapor, is an important active filler in rubber compounding. Its use as a white pigment in paints is decreasing because of the superior properties of tita-

| Table 3.2  Consumption of Fillers | | |
|---|---|---|
| | Consumption, 1000 tons | |
| **Type** | **1989** | **1993** |
| Mineral fillers | 2423 | 2458 |
| Synthetic fillers | 507 | 846 |
| Organic fillers | 97 | 1207 |

nium dioxide. However, it continues to be used as a nontoxic fungicide in paints, as a photoconductive pigment in xerographic paper coatings, and as an ultraviolet stabilizer in plastics.

## Conclusions

The use of properly selected fillers extends the resin supply, and the use of coupling agents with these composites improves performance of plastics and reduces costs and energy consumption. The annual sales of fillers is currently $750 million, and this volume will increase to $1.5 billion during the 1990s.

The end use of fillers in the U.S. is as follows: paper (48%), paint (10%), and plastics (15%). The product distribution is as follows: calcium carbonate (49%), kaolin (18%), silica (14%), and talc (6%). Other authors provide the following data for the annual use of fillers in plastics in the U.S. in tons: asbestos (238,000), talc (75,000), mica (188,000), calcium carbonate (136,000), alumina trihydrate (ATH) (142,000), wood flour (113,000), clay (84,000), silica (50,000), carbon black (31,000), nonfibrous glass (8,000) and cork, pearlite, etc. (3000). A quantity of 660,000 tons of carbon black and an equal amount of noncarbon filler are used annually by the American rubber industry.

According to Business Communications Company, the polymer filler market will approach $3 billion in 1993. Data on consumption of fillers in thousands of tons and anticipated production in 1993 are shown in Table 3.2.

## FIBROUS REINFORCEMENTS

Some of the fillers discussed in this chapter, such as asbestos, have high aspect ratios but are still classified as particulate fillers. Some others, such as microcrystals, are characterized by moderately high aspect ratios but are classified as fibers. These and fibers with high aspect ratios are discussed in Chapter 4.

## REFERENCES

- R.D. Agarwal and L.J. Broutman, *Analysis and Peformance of Fiber Composites*, Wiley-Interscience, New York, 1980
- L. Bohlin *et al.*, *Plast. Comp.*, Vol. 13 (No. 1), 1990, p. 32
- R.D. Deanin and N.R. Schott, *Fillers and Reinforcements for Plastics*, Advances in Chemistry Series 134, American Chemical Society, Washington, DC, 1974
- R. Gachter and H. Muller, Ed., *Plastics Additives Handbook*, Hanser Publishers, New York, 1987
- H.S. Katz and J.V. Milewsky, Ed., *Handbook of Fillers and Reinforcements for Plastics*, Van Nostrand Reinhold, New York, 1978
- J. Kestler, *Plastics Today*, Vol. 1 (No. 14), 1989, p. 7
- G. Kraus, *Reinforcement of Elastomers*, Interscience, New York, 1965
- G. Lubin, Ed., *Handbook of Fiberglass and Advanced Plastic Composites*, Robert E. Krieger Publishing, Huntington, NY, 1975
- G. Lubin, Ed., *Handbook of Composites*, Van Nostrand Reinhold, New York, 1982
- V. Malpass and J.T. Kempthorn, *Plast. Comp.*, Vol. 12 (No. 2), 1989, p. 52
- V. Malpass, J.T. Kempthorn, and A.F. Dean, (Calcium Carbonate), *Plast. Eng.*, Vol. 45 (No. 1), 1989, p. 27
- R.J. Martino, Ed., *Modern Plastics Encyclopedia*, McGraw-Hill, New York, 1986
- J.A. Radosta, *PMAD (SPE)*, Vol. 1 (No. 1), 1984, p. 6
- B. Sanschagrin *et al.*, *Plast. Comp.*, Vol. 10 (No. 3), 1987, p. 37
- D. Sekutowski, *PMAD (SPE)*, Vol. 5 (No. 2), 1987, p. 5
- R.B. Seymour, Chapter 5 in *Developments in Plastic Technology 2*, A. Whelan and J.L. Crafts, Ed., Elsevier Applied Science Publishers, London, 1985
- R.B. Seymour and R.D. Deanin, Ed., *History of Polymeric Composites*, VNU Science Press, Utrecht, Netherlands, 1987
- R.B. Seymour, Ed., *Additives for Plastics*, Vol. I and II, Academic Press, New York, 1978

- R.P. Sheldon, *Composite Polymeric Materials*, Applied Science Publishers, London, 1982
- J.W. Snyder and M.H. Leonard, Chapter 8 in *Introduction to Rubber Technology*, M. Morton, Ed., Reinhold Publishing, New York, 1959
- D.H. Solomon and D.G. Hawthorne, *Chemistry of Pigments and Fillers*, Wiley-Interscience, New York, 1983
- W.V. Titow and B.J. Lanham, *Reinforced Thermoplastics*, John Wiley & Sons, New York, 1975
- A.A. Watts, Ed., *Commercial Opportunities for Advanced Composites*, American Society for Testing and Materials, Philadelphia, 1980
- R.T. Woodhans and I. Chun, *PMAD (SPE)*, Vol. 5 (No. 2), 1987, p. 5

# CHAPTER 4
# Reinforcements

## INTRODUCTION

Natural composites, such as wood, have been available for thousands of years. Ancient artisans recognized the symbiotic relationship between continuous and discontinuous phases when they used pitch to bind reeds to produce composite boats 7000 years ago. Subsequently, in 1500 B.C., artisans in Thebes improved the properties of wood by producing wood veneer.

The continuous phase, used prior to the 20th century, was based on natural resinous products, such as pitch, casein, and albumin. The first synthetic laminating resins was a polyester produced by Berzelius in 1847. This was the precursor of Watson Smith's Glyptals and Baekeland's phenolic resins, which were introduced in the early 1900s. Small amounts of phenolic-based paper and cloth laminates were used for several decades, but the true beginning of the age of composites was the production of fiberglass-reinforced unsaturated polyesters by Ellis and Rust in the late 1930s.

Additives with high aspect ratios, such as asbestos and alpha cellulose as well as glass spheres, were discussed in Chapter 3. Other additives with higher aspect ratios, such as fiberglass and whiskers, will be discussed in this chapter. Since fiberglass is the most widely used and most thoroughly investigated reinforcing fiber, its properties and applications will be emphasized in this chapter.

## FIBERGLASS

The American Society for Testing and Materials (ASTM), in Standard C167-71, defines glass as "an inorganic product of fusion, which has cooled to a rigid condition without crystallizing." Because glass is amorphous, it is isotropic, and like other amorphous polymers has a glass transition point rather than a melting point or first-order transition characteristic of crystalline products.

Sheet glass was used for glazing in 2500 B.C., but fiberglass was not produced until 1620, by Antonio Neri in Florence. Glasslike fibers, called Pele's hair, formed by the wind passing over volcanoes, had been known for many centuries before the fibers were produced by Neri. Nevertheless, these fibers were not available commercially until the 1930s, when Owens Illinois and Corning Glassworks formed Owens-Corning Fiberglas Corporation for the production of fiberglass. Several other firms are now producing this important product as well.

Fiberglass is made from molten glass marbles forced at 1266 °C through orifices in the base of bushings to produce continuous or staple (discontinuous) fibers. The glass is not a definite compound, but is primarily silica produced by heating sand ($SiO_2$), limestone ($CaCO_3$), and boric acid ($H_3BO_3$) in a high-temperature refractory furnace.

Continuous filaments are produced by allowing the molten glass, held in platinum alloy tanks (bushings), to flow by gravity through multiple orifices. The molten filaments formed are gathered together and attenuated to specified dimensions before being quenched by a water spray. The cooled filaments are carried on a belt where they are coated with a lubricant or sizing and grouped together in bundles (strands) which are then wound on spools. The strands are wound together to produce rovings.

Staple fibers are produced by passing a stream of air across the filaments as they emerge from the orifices. The cooled fibers, which are 20 to 40 cm in length, are collected on a rotating drum, sprayed with a sizing, and gathered into strands. Fiberglass mat can be produced by chopping the collection of strands into 2.5 to 5.0 cm long staples. Continuous-strand mat is produced by interlocking the filaments in a spiral fashion. A very thin mat of single filaments is used to produce a surfacing veil for laminated plastics.

In addition to being available as continuous filaments and staple fibers in mats, fiberglass textiles are also available as biaxial, triaxial, knitted, and three-dimensional braided patterns. Many different resin matrices are in use, but the emphasis in this chapter will be on unsaturated polyester and epoxy resins. While strength and stiffness are controlled primarily by the reinforcements, the resinous matrix contributes to thermal conductivity and flexibility. The ultimate properties of these composites are based on

| Table 4.1 Properties of Several Glasses | | | | |
|---|---|---|---|---|
| | | Type of glass | | |
| **Property** | **A** | **C** | **E** | **S/R** |
| Softening point, °C | 727 | 749 | 841 | ... |
| Coefficient of linear expansion, cm/cm/°C | 8.6 | 7.2 | 5.0 | 5.6 |
| Tensile strength, MPa | 3033 | 3033 | 3448 | 4585 |
| % Elongation | ... | 4.8 | 4.8 | 5.7 |
| Specific gravity | 2.5 | 2.5 | 2.55 | 2.5 |
| Index of refraction | ... | ... | 1.547 | 1.523 |
| Note: MPa × 145 = psi | | | | |

a harmonious contribution of both the continuous and discontinuous phases.

A lime-alumina-borosilicate glass called E glass, containing relatively high percentages of alumina ($Al_2O_3$), calcium oxide (CaO), and boric oxide ($B_2O_3$) in addition to silica ($SiO_2$), was developed specifically for the production of fiberglass for electrical (E) applications. This high tensile glass is the major product used as a reinforcement for plastic composites.

High alkali glass (soda or bottle glass), called A glass, is used as a general-purpose reinforcement for composites. Another type of glass called C glass is a sodium borosilicate glass with a high tensile strength used in corrosive environments. Still other commercial types called S glass and R glass have higher tensile strengths and maintain more of their strength at elevated temperatures than E glass. AR glass is an alkali-resistant glass used in the reinforcement of concrete. The properties of these types of glasses are shown in Table 4.1.

There are also a few specialized types of glass fibers, such as type 30, used for continuous pultrusion; S-2 glass, used for filament winding; and leached glass, which is a high-silica glass (Refrasil, Sil Temp, Astrosil, Alphasil). Over 600,000 tons of glass fibers are produced annually in the U.S.

Single filaments (singles yarn) are twisted, 40 times per meter. Heavier yarns are produced by twisting two or more strands together. If the twist resembles the letter S when the yarn is held in a vertical position, it is said to have an "S" twist; if the spirals resemble the letter Z, the singles yarn is said to have a "Z" twist.

The composition of glass is designated by the letters A, C, E, S, and R. A second letter is used to show whether the filament is continuous (C), staple (S), or textured (T). Texturizing or bulking is produced by impinging air on the yarn surface to produce random breakage of the surface filaments (fluffing). A third letter is used to designate the diameter of the yarn: G indicates 9 mm, and P indicates 18 mm. Thus, a yarn designated as ECG is a continuous filament of E glass with a 9 mm diameter.

It is customary to apply a sizing (film), such as polyvinyl acetate (PVAc), and a coupling agent to the fiberglass rovings. As stated in Chapter 3, coupling agents include methacrylato chromic chloride complex (Volan), organosilanes, and organotitanates.

## REINFORCEMENT BY FIBERGLASS

By definition, fibers have a high aspect ratio (HAR). The extent of reinforcement is related to HAR values and orientation, which should be in the direction of the force profile. Since external forces are transferred from the continuous phase (resin) to the discontinuous phase (fiber), optimum adhesion between the two phases is essential for high performance.

Because the law of mixtures is applicable to composites, the modulus or tensile strength of an ideal fiber-reinforced resin composite, consisting of uniformly oriented continuous fibers within a uniform elastic matrix, tested in the direction of orientation of the reinforcement, is equal to the sum of the product of the physical properties and the volume fractions of the two components, i.e., the resin and the reinforcement.

Thus, as shown by the following equation, the modulus of an ideal composite ($E_c$) is equal to the product of the modulus of the fibrous component ($E_f$) and its partial volume ($V_F$) plus the product of the modulus of the resinous matrix ($E_m$) and its partial volume ($1 - V_F$). Other physical properties, such as tensile strength, may be substituted for the modulus.

$$E_c = E_f V_F + E_m(1 - V_f)$$

For less ideal cases, in which the fibers are not uniformly oriented and the test is not made in the direction of orientation, an alignment factor (n) may be introduced to account for fiber disorientation and testing in directions other than that of fiber orientation. The modified equation is:

$$E_c = nE_fV_f + E_m(1 - V_f)$$

While the ideal system of continuous aligned fibers, tested in the direction of orientation, is approached in a pultruded reinforced pipe or profile, most fiber-reinforced composites consist of discontinuous fibers of finite length in which the load on the matrix is transferred to the filler through the effect of shear at the resin-fiber interface. Accordingly, the following equation for the strength of a randomly oriented discontinuous fibrous composite has been proposed, where $\tau_m$ is the shear strength of the composite at its weakest interface, $\xi$ is the strength efficient factor, $\sigma'c$ is the longitudinal strength of the continuous fiber composite, and $\sigma_m$ is the strength of the matrix:

$$\sigma_o = \frac{2\tau_m}{\pi(2 + \ln \varepsilon \ \sigma'c_m/\tau_m^2)}$$

The strength and stiffness of the composite are also related to the glass content. The strength of a fiberglass composite may be increased 400% by increasing the glass content from 20 to 80%. Maximum strength, in the longitudinal direction, is noted when the filaments are all parallel, as in the case of pultrusion. Lower strength is observed when the composite becomes isotropic through a random arrangement of filaments. The maximum glass content present in the composite decreases as the composite changes from anisotropic to isotropic.

The physical properties of glass filament filled composites are dependent on fiber length and vary with direction; e.g., reinforced plastics such as laminates are anisotropic. Differences in the transverse and longitudual directions can be reduced by the use of braided fiberglass or chopped fibers. The effect of the length of 40% filled E glass fibers on the properties of phenolic resin composites is shown in Table 4.2.

## FABRICATION TECHNIQUES

Various fabrication techniques used in the reinforced plastics industry are discussed below.

Table 4.2  Effect of Fiber Length on Properties of
40% E Glass Filled Phenolic Plastics

| Property | 0 | <0.25 | 0.25 | 0.5 | 1 |
|---|---|---|---|---|---|
| | | Fiber length, in. | | | |
| Tensile strength, MPa | 27 | 68 | 10 | 11 | 12 |
| Flexural strength, MPa | 62 | 82 | 12 | 17 | 20 |
| Flexural modified, GPa | 2.4 | 20 | 20 | 20 | 24 |
| Compressive strength, MPa | 8.2 | 24 | 24 | 24 | 24 |

## Hand Lay-Up and Spray-Up Procedures

In one of the simplest and most labor-intensive fabrication procedures, pigmented, catalyzed resin is applied as a gel coat to the surface of the mold. This gel coat, in room-temperature lay-up techniques, ends up on the surface of the finished composite. Catalyzed resin-impregnated mat is then applied over the gel coat; this and subsequent layers are brushed or rolled to ensure good contact between layers and to remove any entrapped air. This procedure is continued until the desired composite thickness is attained.

The assembled composite may be cured at room temperature or at elevated temperatures for faster cycles. This procedure, originally called contact molding, may be upgraded by the application of a vacuum or a pressure bag placed over a cellophane film on the final layer to reduce void formation in the composite. The laminate may also be built up by a spray-up process in which a mixture of chopped glass strands and catalyzed resin is sprayed on the gel coat. In any case, the inner surface will be less smooth than the first layer formed by the gel coat. Tanks, boats, and pipe can be fabricated by this technique. The exposure of workers to styrene has been reduced by using robotic spray-up systems.

## Filament Winding

In a more sophisticated and less labor-intensive procedure, called filament winding, catalyzed resin-impregnated rovings or single strands are wound on a mandrel of appropriate design. The winding machines are programmed to place the resin-impregnated glass in a pattern designed for obtaining optimum properties. The winding is discontinued after the de-

*Fig. 4.1 Filament-wound fiberglass-reinforced epoxy piping system*

sired thickness has been obtained, and the composite is allowed to cure at room temperature or is heated at elevated temperatures. Figure 4.1 shows a system of epoxy pipes produced by filament winding.

## Centrifugal Casting

Fiber-reinforced plastic pipe can be produced by rotating a mixture of chopped strand and catalyzed resin inside a hollow mandrel. Because of differences in specific gravity, there is a tendency for these composites to be less homogeneous than those produced by other techniques.

## Pultrusion

In a modification of the thermoplastic extrusion technique, continuous resin-impregnated strands are drawn through a die which shapes the extrudate and controls the resin content. The composite is then drawn through a curing oven. Pipe and sheet can be produced continuously by the pultrusion technique.

Pultrez Ltd. (U.K.), which has 70 pultrusion lines in 18 different countries, is integrating a pull winding technique with a pultrusion process. Goldsworthy, who pioneered pultrusion and filament winding processes, has discussed a system that combines tape placement and filament winding technology for the construction of unique large structures. It has been predicted that pultrusion will be a $5 billion business in 1995.

## Matched Die Molding

Matched die molding of a premix of chopped glass, roving, and catalyzed resin is used for relatively large-scale production of reinforced articles. Uncured, doughlike compositions are called bulk molding compounds (BMC). Uncured resin-impregnated sheets are called sheet molding compounds (SMC). These compounds are supplemented by thick molding compounds (TMC) and XMC. TMC is produced continuously on a machine that resembles a two-roll mill. XMC, in which the continuous impregnated fibers are arranged in an X-pattern, is produced on a filament winding machine.

SMC may be produced continuously by the melt impregnation process, in which a preheated reinforcement sheet is impregnated on both top and bottom by an extruded prepolymer melt. Two or more impregnated mats are passed through a rotating double belt to form the prepolymer sheet. SMC can also be produced by the deposition of an aqueous slurry of chopped fibers and a powered resin on a moving screen prior to drying and preheating of the composite.

Computer-aided mold analysis, advanced press technology, faster curing materials, and more automated secondary operations will be essential for SMC production in the future. Microcomputer controls for high-speed, high-tech compression presses for SMC and glass-reinforced thermoplastic stampable sheet are also essential for the production of fiber-reinforced plastics. One of the newest inventions is the automated "camphor" process, which permits the addition of ribs, closed sections, core, and encapsulation of other materials.

Liquid composite molding (LCM), which embraces resin transfer molding (RTM) and reaction injection molding (RIM), ensures the production of quality high-volume composite parts.

## Injection Molding

A composite premix similar to BMC can be injection molded in much the same manner as thermoplastics. Of course, both thermoset and thermoplastic composites can be injection molded.

## BASALT

Because basalt is igneous rock formed by cooling lava, it can be remelted. Major plateaus of basalt occur in Washington state near the Columbia River, and in the Deccan Trap, Siberian Trap, Piranhas Basin, Karroo, and Tasmania. Basalt fibers have produced commercially in the U.S.S.R. and Germany and experimentally in Washington state.

## ARAMIDS

Aramid fibers (Kevlar), produced by the condensation of terephthaloyl chloride ($ClOCC_6H_4COCl$) and $p$-phenylenediamine ($H_2N-C_6H_4-NH_2$), have been available commercially since 1971. The strength of these rigid macromolecules in aramids is enhanced by intermolecular hydrogen bonding. Because of their rodlike structure, aramids have exceptionally high modulus (stiffness) and high packing efficiency. These crystalline fibers are more flexible than glass fibers. High-melting fibers such as Kevlar 29 and Kevlar 49, which are spun from concentrated sulfuric acid using techniques described for fiberglass, are being used in resinous composites. These fibers have a low specific gravity of 1.44.

Aramids have a negative coefficient of expansion in the fiber direction and a positive coefficient in the radial directions of the fiber. Aramid fibers have high thermal stability, low dielectric properties (nonconductors), and good resistance to chemicals, except strong acids and alkalies. They may be used at temperatures up to 200 °C. The properties of commercial aramid fibers are shown in Table 4.3.

## CERAMIC FIBERS

Ceramic fibers (nonmetallic, inorganic fibers) include alumina-silica, alumina-silicon carbide, and alumina-boria-silica fibers. Refractory Fiberfrax ceramic fibers (RFC, Fiberfax) are alumina-silica fibers that have been commercially available since the early 1960s. These heat-resistant fibers are available as staple, chopped, and milled fibers.

| Table 4.3  Properties of Commercial Aramid Fibers | | |
|---|---|---|
| Fiber type | Tensile strength, GPa | Tensile modulus, GPa |
| 29 | 3.6 | 83 |
| 49 | 3.6 | 124 |
| 149 | 3.4 | 172 |
| 68 | 3.6 | 110 |
| 129 | 4.2 | 110 |

| Table 4.4  Properties of Typical Ceramic Fibers | | | |
|---|---|---|---|
| Fiber type | Tensile strength, GPa | Modulus, GPa | Specific gravity |
| Alumina-silica | 1.7 | 93 | 2.70 |
| Silicon carbide | 2.8 | 193 | 2.56 |
| Alumina-boria-silica | 1.7 | 151 | 2.50 |
| Alumina | 1.4 | 379 | 3.90 |

High-modulus, high-melting fibers (Fiber FP, Saffil) are being produced commercially from alpha alumina. This fiber is supplied in small diameters (15 to 25 nm) and retains much of its strength at 1000 °C. Its melting point is 2045 °C.

Silicon carbide, which is produced by the pyrolysis of rice hulls, is available as continuous filaments and staple fibers. Alumina-boria-silica (Nextel) fibers are also available as continuous filaments and staple fibers. These fibers retain their good physical properties at temperatures as high as 1250 °C. The properties of commercial ceramic fibers are shown in Table 4.4.

# POLYBENZIMIDAZOLE

Polybenzimidazole (PBI) fibers, which are produced by the condensation of diphenyl isophthalate ($(C_6H_5OOC)_2C_6H_4$) and $3,3',4,4'$-tetraamino-biphenyl ($(H_2N)_2C_6H_4$-$C_6H_4(NH_2)_2$), have a high melting point and are resistant to flame. PBI has been used experimentally as a reinforcing fiber in resin composites in aircraft applications.

# BORON

Boron filaments (Borofil) are produced by deposition from a boron trichloride $BCl_3$-$H_2$ mixture on a fine tungsten filament, similar to that used in incandescent lamps. Boron may also be deposited on a carbon filament. Since these high-strength boron filaments are expensive, their use has been limited to unique applications, such as helicopter blades and empennage structures for F-14 and F-15 aircraft. The tensile strength and modulus of boron fibers are 3.5 and 380 GPa, respectively.

# CARBON (GRAPHITE)

Carbon fibers, produced by the pyrolysis of acrylic filaments (PAN), are the predominant high-strength, high-modulus fibers used as reinforcements for polymeric composites. The principal disadvantage of these low specific gravity (1.8) fibers is their low strain to fracture property.

The term "carbon fibers" is used for fibers processed at temperatures below 1700 °C and with a tensile strength less than 345 GPa. Graphite fibers, which are heat treated at higher temperatures (above 1700 °C), have tensile moduli of at least 345 GPa and a high degree of orientation. Carbon fibers are also produced by melt spinning isotropic pitch and carbonizing and stress-graphitizing the product at higher temperatures. About 3000 tons of carbon-graphite fibers are produced annually in the U.S.

The Lockheed F-117A fighter and the Northrop B-2 bomber, like all composites, are radar transparent. Space Shuttle doors and antennae are being constructed from graphite-reinforced epoxy resins, and plastic composites have replaced aluminum in short takeoff U.S. Marine Corps Hammer jet-propelled vehicles.

The Beech Starship, which was designed by B. Rutan, consisted of 2600 composite components instead of the alternative 10,000 metal components. A five-seat helicopter that functions without a tail rotor has been

| Table 4.5 Physical Properties of Typical Carbon Fibers | | | |
|---|---|---|---|
| **Fiber type** | **Tensile strength, GPa** | **Modulus, GPa** | **Source** |
| Low modulus | 0.34-100 | 34-69 | Rayon, pitch |
| Intermediate modulus | 2.76-5.50 | 136-275 | PAN |
| High modulus | 2.07-2.76 | 344-414 | PAN, pitch |
| Very high modulus | 1.13-2.4 | 483-690 | PAN, pitch |

| Table 4.6 U.S. Producers of Carbon Fibers | | |
|---|---|---|
| | **Capacity, tons** | |
| **Company** | **Fiber from PAN** | **Fiber from pitch** |
| Hercules | 1250 | ... |
| Union Carbide | 400 | 100 |
| CCF | 500 | ... |
| Great Lakes | 500 | ... |
| Hysol Grafil | 550 | ... |
| Ashland (Carboflex) | 150 | ... |
| Hitco | 100 | ... |
| Stackfield | 75 | ... |
| Avco | 40 | ... |

constructed with a composite beam stabilizer. The nozzle cones for solid rocket motors are made from carbonized rayon filament-reinforced epoxy resins. However, because of environmental problems, production of these carbon filaments has been discontinued, and filaments produced by the pyrolysis of PAN cannot be used for this composite application. The physical properties of carbon fibers are shown in Table 4.5. American producers of carbon fibers and their plant capacities are shown in Table 4.6.

## SILICA-ZIRCONIA

As stated in the discussion of fiberglass, high-silica filaments are being produced commercially by hot acid leaching of fiberglass. Quartz fibers (Astroquartz and Alphaquartz), which have much higher tensile strengths than fiberglass, are used in composites with or without resin-compatible finishes. Aluminum silicate (Fiberfax), boron nitride, aluminum borosilica (Nextal), and silicon carbide fibers (Silag) are also available. Zirconium oxide fibers (Zircar) are also useful reinforcements for high tensile service. Nextal fibers are being used primarily for ceramic reinforcements. A new family of heterocyclic rigid rod and chain-extended polymeric fibers (PBZ) are being used in structural composites. Polyimide films are being used, to a limited extent, in composites. Polyphenylene sulfide fibers (Sulfar) and silicon carbide fibers are also available.

## OTHER ORGANIC FIBERS

Experimental composites have been produced from liquid crystal polyesters (Vectran) and high-modulus polyesters (Compet). Extremely strong oriented extended-chain polyolefin filaments (Spectra) are being produced by gel spinning of the resin in hot paraffin and trichlorotrifluorethane. Despite their lack of high-temperature resistance, these fibers are being used experimentally for reinforcing polymer composites.

The poor adhesion of thermoplastics to fiber reinforcements has been overcome, to a large extent, by the use of fibers produced by blending thermotropic liquid crystalline polymers (LCP) with engineering plastics to generate a fibril morphology in the LCP phase. The modulus of the drawn film, consisting of 80% of polyphenylene sulfide (PPS) and 20% of an LCP, based on a copolymer of 70% *p*-hydroxybenzoic acid (PHB) and 30% polyethylene terephthalate (PET), increased from 6.4 to 25.1 GPa as the draw ratio was increased from 1.9 to 25.1.

The tensile strength of crystalline reinforced thermoplastics (RTP) is much greater than that of amorphous RTP's. For example, the tensile strength of nylon is increased as much as 150%, whereas the tensile strength of acrylonitrile-butadiene-styrene (ABS) is increased only 75% by reinforcement with fiberglass.

## REINFORCEMENT BY SHORT FIBERS

The theories developed for predicting the physical properties of fiber-reinforced composites may be used, with modifications, for composites with short fibers and particulate fillers. In this modification, one defines a critical fiber length ($L_c$) as the minimum length per given fiber diameter essential for high tensile fracture stress. At fiber lengths below $L_c$, the fibers are not long enough to be effective if adhered to the resin matrix, and slip and are pulled out under tensile loadings. The critical fiber length ($L_c$) is directly proportional to the diameter (D) and tensile strength of the fiber at the break point ($T_f$), and inversely proportional to the adhesive bond between the matrix and the fiber surface (A); the latter may be assumed to be equal to the strength of the resin:

$$L_c = \frac{DT_f}{2A}$$

Since a distribution of fibers of subcritical lengths (i.e., below $L_c$) is typical, the composite strength $\sigma_c$ is given by the complex equation:

$$\sigma_c = C\left[\sum \frac{\tau L_i V_i}{2r} + \sum E_f e_c \left(1 - \frac{E_f e_c r}{2L_j \tau}\right)V_j\right] + E_m e_c (1 - V_f)$$

where C is the orientation factor, $\tau$ is the shear strength of the interfacial bond, r is the radius of the fiber, $L_i$ is the fiber length for the subcritical fibers, $V_i$ is the volume fraction of the subcritical fibers, $V_f$ is the total volume fraction of the fibers, $e_c$ is the strain, and $E_m$ and $E_f$ are the moduli of the matrix resin and fiber, respectively.

The optimum length for glass and graphite reinforcing fibers is 1 to 2 mm, and the optimum diameter of glass fibers is 0.005 mm; the optimum fiber matrix modulus ratio is 50/1.

Adequate interfacial bonding is approached when coupling agents, with selected functional groups for fiber and resin, are present. However, an efficiency factor $K_{eff}$ is usually incorporated in the previously cited additive rule of mixtures:

$$E_c = k_{eff}E_f V_f + E_m(1 - V_f)$$

## WHISKERS

Small single crystalline filaments were originally produced from gamma alumina (sapphire), but much of the interest today is focused on silicon carbide (SiC) whiskers. Most whiskers are produced by vapor phase reactions of iron, copper, chromium, silicon, silicon carbide, tin, and zinc. Silicon carbide whiskers are produced by the pyrolysis of rice hulls.

## MICROFIBERS

Potassium titanate microfibers (Fybex) were produced originally by DuPont by heating titanium dioxide, potassium molybdate, and potassium carbonate. DuPont has discontinued the production of potassium titanate, but this fiber is now being produced by Otsuko Chemical in Japan at an annual rate of 50 tons. This dark blue-colored filler has an aspect ratio of 40/1 and an index of refraction of 2.35. Sodium aluminum hydroxy-carbonate microfibers (Dawsonite; $NaAl(OH)_2CO_3$) are produced by the Aluminum Company of America (Alcoa). This filler has an aspect ratio of at least 25/1 and an index of refraction of 1.53. It undergoes a sharp weight loss as a result of the evolution of steam and carbon dioxide at 300 °C. It is a flame retardant and smoke depressant as well as an HCl scavenger. However, preliminary tests have indicated that acicular Dawsonite filler is a carcinogen.

Franklin fiber is a β-anhydrite, white calcium sulphate microfiber produced by Certain-Teed Products Corporation. This moderately priced microfiber has an aspect ratio of at least 25, a specific gravity of 3, and an index of refraction of 1.58. Preliminary tests show it to be nontoxic.

Processed mineral fiber (PMF) is primarily a calcium silicate blast furnace slag. This light grey microfiber has an aspect ratio of 40/1 and a specific gravity of 2.7. Calcium aluminum silicate is also being used experimentally as a reinforcing microfiber.

## NATURALLY OCCURRING
## ORGANIC FIBROUS MATERIALS

In addition to attrition-ground wood flour, which has some fibrous properties, other ground waste products with less fibrous characteristics, such as cork flour, lignin, ground corncobs, and ground nut shells, have been used to a small extent as fillers. Natural fibers such as α-cellulose, cotton linters, sisal, and jute have also been used on a small scale. These additives degrade at processing temperatures above 180 °C and have impact strengths inferior to those of longer fiber additives.

## POLYESTER AND POLYAMIDE FIBERS

Polyethylene terephthalate (PET) and nylon fibers improve the impact strength of composites and are sometimes admixed with other fibers, such as glass, and used as hybrid fibers.

## HYBRID FIBERS

Since aramid and carbon fibers have similar coefficients of expansion, they are sometimes used as hybrid fibrous additives in composites. Aramid fiber composites are characterized by a lower compressive strength and higher impact resistance than carbon fiber composites. Hence, the hybrids represent a compromise.

Additional improvements are observed when the composite consists of low specific gravity and high-modulus aramid and carbon fibers along with higher specific gravity fiberglass. Aramid and glass hybrid fibers are used to reinforce epoxy resins in boats. Relatively high impact-resistant composites for use in aircraft are produced by reinforcing with carbon/glass hybrids.

## POLYVINYL ALCOHOL

Polyvinyl alcohol (PVA) fibers, insolubilized by surface treatment with formaldehyde, provide composites with good flexibility, low specific gravity, and high impact strength. PVA-reinforced polyesters have been used as railroad ties in Japan.

## POLYMERIC FOAMS

Cellular plastics can be stabilized by additives and reinforced by fillers or fibers. Nevertheless, foams without these additives consist of resins and gas and, hence, are classified as composites. These foams will be discussed in Chapter 5.

*Fig. 4.2 Reinforced polyester boat (Amoco Chemicals)*

## FUTURE OUTLOOK

Of the 1745 boats exhibited at the 1989 Genoa boat show, more than 97.5% were based on polyester and epoxy resins reinforced with glass, carbon, and aramid fibers (see Fig. 4.2). Advanced composites are also widely used for nautical equipment and in the construction of high-performance yachts.

Reinforced polyphenylene sulfide (PPS) is being used for the fabrication of strong knee braces. According to Business Communication Company, a large percentage of the 708 million lb of RTP's used in the U.S. in 1988 and the 128 billion lb of RTP's to be used in 1995 are consumed by the automotive field.

Business Communication Company projects annual growth of 7.7% for reinforced thermosets. According to Goldsworthy, the composites structure industry is changing rapidly from a specialty field to one in which high-volume composites are produced by automated fabrication. Courtaulds predicts a threefold increase in the European demand and a worldwide demand of 13,300 tons for carbon fibers by 1995. 1.35 million lb of composites were produced by the pultrusion technique in 1988. As stated in an advertisement by Himont, the *Spruce Goose* might really have gotten off the ground if Howard Hughes had used a fiberglass-reinforced resin composite instead of wood in the construction of his eight-propeller aircraft now on exhibit at Long Beach, CA. Because of the availability of advanced plastic composites, the sky is no longer the limit with modern aircraft.

## REFERENCES

- D.M. Bigg, SMC Sheet Production, *Plast Eng.*, Vol. 46 (No. 10), 1990, p. 33
- J. Delmonte, *Technology of Carbon and Graphite Fiber Composites*, Van Nostrand Reinhold, New York, 1981
- E. Doll, (Epoxy resin composites), *Modern Plastics*, Vol. 66 (No. 2), 1989, p. 77
- U.N. Epel, J.M. Margolis, S. Neuman, and R.B. Seymour, Ed., *Engineered Materials Handbook*, Vol. 2, *Engineering Plastics*, ASM International, Metals Park, OH, 1988
- K.M. Finlayson, (Carbon fibers), Tectronic Publishing, Lancaster, PA, 1989
- W.B. Goldsworthy, (Future outlook), *Plastics Today*, Vol. 1 (No. 2), 1989, p. 14
- N.L. Hancox, Ed., *Fiber Composite Hybrid Materials*, Macmillan, New York, 1981
- G.S. Kirshenbaum, (Composite education), *Polymer News*, Vol. 14 (No. 37), 1989
- M.W. Liedtke and W.H. Todd, Ed., *Advanced Composites*, ASM International, Metals Park, OH, 1985
- G. Lubin, Ed., *Handbook of Composites*, Van Nostrand, New York, 1982
- M.H. Naitove, (Liquid-composite molding), *Plast. Technol.*, Vol. 35 (No. 3), 1989, p. 50
- F.J. Phillips, (Carbon fibers), *Modern Plastics*, Vol. 66 (No. 11), 1989, p. 232
- J.J. Pigliacampl and P.G. Rienwald, Aramid Fibers, *Modern Plastics*, Vol. 66 (No. 11), 1989, p. 230
- T.J. Reinhart, Ed., *Engineered Materials Handbook*, Vol. 1, *Composites*, ASM International, Metals Park, OH, 1987
- J.K. Rogers, (LCM), *Plast. Technol.*, Vol. 35 (No. 3), 1989, p. 51
- J.K. Rogers, (SMC), *Plast. Technol.*, Vol. 34 (No. 10), 1988, p. 47

- J.K. Rogers, (Auto Producers Consortium), *Plast. Technol.*, Vol. 35 (No. 10), 1989, p. 108
- P.G. Rose, (Short fiber-reinforced plastics), *Pop. Plastics*, Vol. 30, (No. 5), 1987, p. 43
- E.P. Rossa, (E glass), *Plast. Eng.*, Vol. 44 (No. 11), 1988, p. 39
- M.M. Schwartz, Ed., *Composite Materials Handbook*, McGraw-Hill, New York, 1984
- R.B. Seymour, *Engineering Polymers Sourcebook*, McGraw-Hill, New York, 1989
- R.B. Seymour, *Polymers for Engineering Applications*, ASM International, Metals Park, OH, 1987
- R.B. Seymour and R.D. Deanin, Ed., *History of Polymer Composites*, VNU Science Press, Utrecht, The Netherlands, 1987
- A.M. Thayer, (Advanced composites), *Chem. Eng. News*, Vol. 68 (No. 30), 1990, p. 37
- W.V. Titow and B.J. Lankam, *Reinforced Thermoplastics*, John Wiley & Sons, New York, 1975
- P.A. Toensmeier, (Reinforced thermoplastics), *Modern Plastics*, Vol. 67 (No. 3), 1990, p. 12
- P.A. Toensmeier, (Robotic spray-up), *Modern Plastics*, Vol. 66 (No. 3), 1989, p. 47
- J.W. Weeton, D.M. Peters, and K.L. Thomas, Ed., *Engineers' Guide to Composite Materials*, ASM International, Metals Park, OH, 1987
- R.J. Wilder, (Pultrusion), *Modern Plastics*, Vol. 65 (No. 7), 1988, p. 37
- M.J. Wirtner, (Ceramic fibers), *Modern Plastics*, Vol. 66 (No. 11), 1989, p. 255

# CHAPTER 5

# Cellular Plastics

## INTRODUCTION

While some technologists do not consider plastic foams to be composites, these materials do consist of a solid resinous phase and a gas phase. The gas phase is actually a lightweight extender.

Of course, natural sponges such as *Hippospongia equina* and *Euspongia officinalis*, which consist of strands of sponges, have been available for centuries. Man-made sponges were not available until Skidrowitz patented a rubber "Quala" sponge in 1914. These sponges, which were formed by the use of a carbon dioxide propellant from the thermal decomposition of sodium carbonate ($Na_2CO_3$) or ammonium carbonate (($NH_4)_2CO_3$), were displaced in the early 1930s by foamed rubber latex. Subsequently, carboxylated butadiene-styrene lattices were used for the production of foams for the backing of tufted carpets.

In addition to these pioneer flexible foams, rigid foams were made from hard rubber (ebonite) in the early 1920s, and extruded foamed polystyrene was produced in Sweden in 1931. A comparable product called "Styrofoam" was produced by Dow in 1943. Expandable solvent-containing polystyrene granules were produced by BASF in the 1950s. Syntactic foams, consisting of a plastic filled with hollow microspheres, were commercialized in 1953. More than 2 million tons of cellular plastics are consumed annually in the U.S. Most of these foams are based on polyurethane and polystyrene, but any solid polymer can be used as a foam. Foams have been made from almost every available commercial plastic and elastomer.

## FOAM FORMATION THEORY

Cellular plastics, which consist of closed or open cells, may be flexible or rigid and are available in a wide range of densities. The gaseous component in some lattices and solutions is introduced by the whipping of air or the vaporization of water; however, most foams are produced by the introduction of a physical (PBA) or chemical blowing agent (CBA). These additives are described in Chapter 2.

There are three successive stages involved in the production of foamed polymers: the liquid stage, the gas bubbling stage, and the solidification stage. If the polymer is present in a latex or solution, it may be expanded into a foam. Thermosets must be expanded before they are crosslinked, and solid thermoplastics must be heated before the introduction of the propellant.

As stated in Chapter 2, the propellant or blowing agent may be a volatile liquid, such as pentane in polystyrene, or a gas produced by the decomposition of a CBA, such as azobisformamide (ABFA). The solidification or stabilization of the expanded polymer may take place by cooling, gelation, or solvent evaporation.

In the expansion process, the growth of the cell is controlled by the difference in external and internal pressure ($\Delta P$), the surface tension ($\gamma$), and the radius of the cell (r), as shown in the following equation:

$$\Delta P = \frac{2\gamma}{r}$$

$\Delta P$, which is inversely related to the difference in the radii of the cells ($r_1$, $r_2$), may be reduced to zero by breakage of the cell wall or by diffusion of the propellant.

$$\Delta P = 2\gamma \left[ \left( \frac{1}{r_2} \right) - \left( \frac{1}{r_1} \right) \right]$$

The foam, at any stage in its development, and the area (A) at any temperature (T), may be described by the ideal gas equation:

$$PV + 2/3 \, \gamma A = nRT$$

The free energy ($\Delta G$) of the system is also related to the second term in the equation of state, i.e.:

$$\Delta G = \gamma A$$

Nucleation of the gas cells is aided by lowering the surface tension ($\gamma$) or by adding silicone oil or another surface-active agent. However, the concentration of the surface-active agent at the gas-polymer interface decreases as the cells expand. This effect may be counteracted by the diffusion of a surface-active agent from the interior to the cell surface, which is called the "Gibbs effect," and by the surface layer flowing from lower to higher areas of surface tension, called the "Marangoni effect." The optimum condition is when the latter is the dominant effect. Van der Waals' forces between surfaces may promote thinning of the cell walls.

## FOAM PRODUCTION

Foam production may take place by expandable formulations, decompression, expansion, or dispersion processes. Polystyrene foam is produced by the mixing of a polymer and a volatile liquid or gas, such as pentane, in a single or twin screw extruder, followed by cooling of the melt and extrusion of the composite through a die. Expandable polystyrene (EPS) beads, obtained by the polymerization of monomer with a blowing agent, may also be extruded to produce foam. However, most EPS is molded from PS beads containing volatile liquids.

The EPS beads may be expanded from 200 to 500% of their original volume by controlling the time and temperature of the molding process. It is customary to use steam or hot air to preexpand EPS to the desired density before molding. Moldings or blocks of EPS are produced in a similar manner. This expansion process has also been used with styrene copolymers, such as styrene-acrylonitrile (SAN), or copolymers of styrene and ethylene, polyethylene, and polypropylene.

The dispersion process is a modification of the cream whipping or meringue process used in cooking. For example, a gas is dispersed in rubber latex, the rubber particles are coalesced by the addition of an acid former, such as sodium silicofluoride ($Na_2SiF_6$), and the expanded matrix is cured and dried.

Open-cell cellular rubber is produced using a chemical blowing agent so that the gas may be released when the rubber solution has a minimum viscosity. Closed-cell cellular rubber is produced when the gas is released in a high viscosity or partly cured polymeric system. Most polymers can be produced with either open (reticular) or closed cells (nonreticular). Dimensional stability can be obtained by crosslinking the cellular product. Foams for rug and floor backings are obtained by blowing polyvinyl chloride (PVC) plastisols. A plastisol is a suspension of PVC in a liquid plasticizer.

## PROPERTIES OF CELLULAR PLASTICS
## (HIGHER DENSITY)

Smaller cells (higher density) are produced when many nuclei are present. Any finely divided substance, such as fumed silica, titanium dioxide, or carbon black, will serve as a nucleating agent. Large bubbles contribute to low density and poorer physical properties. Low-density foams are preferred for buoyancy applications, but open cells are preferred for acoustical applications.

In general, the properties of cellular polymers are related to those of the solid polymer precursor. However, the foams are less rigid and the extent of softening is related directly to the cell size. The chemical and flame resistance of foams is similar to those properties in the solid polymer, but the higher surface/volume ratio in foams enhances flammability. Foams produced by hydrocarbon propellants are more flammable than those produced by chlorofluorocarbon (CFC) blowing agents, but the use of the latter has been discontinued because of their adverse effect on the ozone layer.

The foaming of polymers lowers physical properties, but enhances thermal, electrical, and insulating properties. The properties of a foam (P), in general, are proportional to the density (D) to the nth power. The proportionality factor and the exponent are related to the specific polymer and the property under consideration.

Properties such as modulus, creep, hysteresis, dimensional stability, and thermal conductivity are in accord with the following equation:

$$P = KD^n$$

The values of K and n for compressive strength in polyurethane foams are 12.8 and 1.54 and the values for flexural strength are 19.0 and 4.36, respectively. Conversely, buoyancy, softness, resilience, shock absorbency, thermal electrical insulation, permeability, and absorption of liquids are inversely related to D.

Strength, elongation, resistance to compression, and thermal insulation are related inversely to cell size, but modulus, shock resistance, and permeability are related directly to cell size. Because expansion usually occurs in a specific direction, many cellular polymers are anisotropic. Thus, compressive modulus and strength would be greater in the direction in which the cells rose during formation. Fillers may be added to cellular polymers to control density and to improve cushioning properties.

## STRUCTURAL FOAMS

The term "structural foam" was coined by R. Angell of Union Carbide in the 1960s to describe a "plastic product having a cellular core, integral skin, and a high enough strength to weight ratio to be called structural." Large structural foams can be readily molded. These moldings have a strength to weight ratio three to four times that of the solid plastics.

The molding of a structural foam is similar to that of the molding of classic foams. However, the first layer of polymer and gas bubbles form a skin as they hit and break against the chilled walls of the mold cavity. Because of its low thermal conductivity, the cellular product must remain in the mold until the skin is strong enough to withstand the continued expansion in the interior of the molded part.

The classic structural foam was made from a mixture of CBA and polystyrene, but nitrogen was substituted for CBA as processing techniques were improved. It is now customary to use a low-pressure multinozzle and to inject the nitrogen partway down the extruder barrel at a point where several flights of the screw are removed in order to develop a significant pressure drop. The polymer-nitrogen mixture is then injected through a manifold system of synchronized nozzles. High-pressure single nozzles, which permit faster rates, are being used in Europe. The nitrogen-polystyrene mixture is injected into a steel mold cavity at extremely high pressures, and the cellular product is allowed to expand hydraulically against the skin.

Melt-processible structural foam is produced in conventional injection molding presses, in which the expansion of the core is controlled by the amount of blowing agent and the level of depressurization. The wall thickness of these structural foams is increased in conjunction with a decrease in density. The term "multicomponent liquid foam processing" is used to describe the molding of an expandable thermoset which is crosslinked in the mold cavity. Modern office furniture made from polyurethane structural foam is shown in Fig. 5.1.

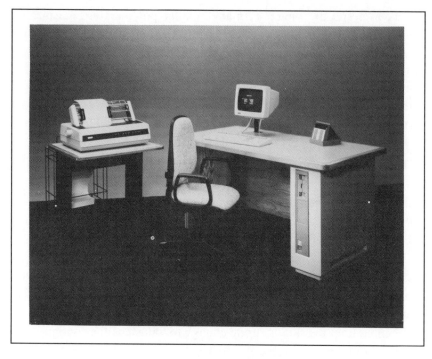

*Fig. 5.1 Modern office equipment produced from polyurethane structural foam by reaction injection molding. (Mobay Chemical Corp.)*

## RIGID FOAMS

The rigid polyurethane foams accidentally discovered by Nobel Laureate O. Bayer in the late 1930s are now produced at an annual rate of 350,000 tons in the U.S. and at a higher rate in Europe and Japan. These and other rigid foams will be described in this section.

### Fluorocarbon Blowing Agents

Annual U.S. consumption of CFC's grew to 305,000 tons in 1986. CFC's were used for foams (28%), refrigeration and air conditioning (40%), cleaning agents (20%), and aerosols (12%). The principal U.S. producers in 1989, with EPA annual allowances in thousand of tons, were: DuPont (152), Allied-Signal (77), Pennwalt (39), LaRoche (28), and Racon (14).

Because the ozone layer that shields the earth from excessive solar radiation is affected by CFC's, foam producers and environmentalists are concerned about the exhaustion of long-lived CFC's into the atmosphere. Under normal circumstances, the rates of formation of the ozone and consumption of the ozone in the ozonosphere are equal; however, free radicals produced by the dissociation of some CFC's decrease the concentration of ozone and hence permit an increase in UV radiation on earth.

Habitation on the planet earth is dependent on the "greenhouse effect," based to a large extent on the presence of atmospheric carbon dioxide, which absorbs the infrared radiation that warms the earth's surface by as much as 35 °C. Since CFC's also contribute to the greenhouse effect, there is reason for concern. J. Hansen, director of the Goddard Institute for Space Studies (GISS), maintains that a "greenhouse warming" is now occurring, as demonstrated by recordbreaking heat spells in the 1980s. Since the greenhouse effect and atmospheric CFC's have a global impact, attempts were made to regulate the evaporation of CFC's into the atmosphere at an international meeting that resulted in the "Montreal Protocol."

Ratification of this Protocol by at least 11 countries, which account for two-thirds of the world's consumption of CFC's, was required before the rules became effective. The Montreal Protocol requires that producers freeze their production of CFC's at 1986 levels and that CFC production be reduced to 50% of the 1986 level by 1998. Since the use of CFC's increased 15% from 1986 to 1989, this edict required an immediate reduction of 15% in CFC production. Additional 20% and 30% reductions in CFC production will be mandated in mid-1993 and 1998, respectively. While the Protocol overlooks the principal source of chlorine in the stratosphere, it mandates significant cuts in the production of CFC-11, CFC-12, CFC-113, CFC-114, and CFC-115 and freezes the production rate of Halon 1211 ($CBrClF_2$), Halon 1301 ($CBrF_3$), and Halon 2402 ($C_2Br_2F_4$).

U.S. Senate and House bills called for taxes on CFC's 11, 12, 113, 114, and 115 and Halons 1201, 1301, and 2402. The bill passed by the U.S. Congress has the following graduated scale: $3.01/kg in 1990, $3.67/kg in 1992, and $6.82/kg thereafter. Senator J. Chafee also introduced amendments which would phase out U.S. production of hydrofluorocarbons (HCFC) by the year 2030.

HCFC's, which cause less than 2% of the CFC ozone depletion, are preferred substitutes for CFC's. Uniroyal is investigating fully halogenated CFC's vs. HCFC's. DuPont has proposed the use of a blend of HCFC 141b and HCFC-123 (Fornil R) as a propellant for making plastic foams. DuPont has also patented the production of HCFC 124, HCFC 123, and blends HCFC 124, 22, and 152a. The effect of these blends on the ozone layer is less than 3% of that of CFC's. DuPont has pledged to abandon CFC production by the year 2000 to meet the requirements of the United Nations environmental program "UNEP."

There is a renewed interest in the classic use of water for the production of carbon dioxide as a blowing agent for polyurethane foams. Suppliers now recognize that reformulations are essential and have moved from the defensive to the offensive on polyurethane foam propellants.

Union Carbide has introduced a system called Geolite based on the use of water and methylene chloride. as well as Hyperlite for auto seating and Ultracil, which reduces volatile organic compound (VOC) emissions. These high-performance, flame-resistant foam systems are said to be based on "soft blowing agents."

## Polyurethane

Polyurethane (PUR) foams were discovered serendipitously by A. Bayer in the late 1930s when a trace of water reacted with alkylene diisocyanate to produce carbon dioxide, resulting in a cellular product which Bayer's colleagues derisively called "imitation Swiss cheese." PUR foams based on hydroxyl-terminated polyesters were introduced commercially by Mobay Chemical Company in 1954. U.S. consumption of flexible and rigid PUR foams in 1989 was 786,000 and 392,000 tons, respectively. Worldwide consumption of PUR's in 1989 was 4.14 million tons. Rigid PUR foam is used for laminated board stock, moldings, foamed-in-place insulation, sprayed foam, and packaging.

The classic commercial rigid polyurethane foams were produced by the reaction of an aromatic diisocyanate such as tolylene diisocyanate ($CH_3C_6H_3(CNO)_2$) (TDI) and an aliphatic polyester with terminal hydroxyl groups (telechelic ester) (HOR'OH). The formation of polyurethane is shown in the following equation:

$$HOR'OH + OCN-Ar-OCN \rightarrow -OR'OOC-NHAr-NH-OOCR'-$$

As shown by the following equation, any water or organic acid present in the PUR reactants will yield carbon dioxide ($CO_2$); this was the blowing agent responsible for the "imitation Swiss cheese" produced by Bayer. Monofunctional reactants are used for the sake of simplicity in the following equations, which show the chemical reaction of isocyanates with water or an organic acid (RCOOH):

$$ArNCO + HOH \rightarrow [ArNHCOOH] \xrightarrow{ArNCO} ArNHCONHR + CO_2$$

$$ArNCO + R''COOH \rightarrow [ArNH-COOCR''] \rightarrow ArNHCOR'' + CO_2$$

Carbon dioxide is still used to some extent today, but most PUR foams are produced by the use of physical blowing agents such as fluorocarbons or other chemical blowing agents. Because of inferior hydrolytic stability, the telechelic polyesters have been replaced by telechelic polyethers, i.e., polyethers with terminal hydroxyl groups.

As stated in Chapter 2, organotin compounds and surfactants are added to the reactants as catalysts and surface tension control agents, respectively. Rigid PUR foam has an insulation value (K factor) twice as high as any other insulating material. The K factor is expressed as $Btu/ft^2 \cdot h \cdot °F$. The Btu (British thermal unit) is a unit on the $ft \cdot lb \cdot s$ scale comparable to the calorie on the metric scale, i.e., the quantity of heat required to raise the temperature of 1 lb of water 1 °F (1 Btu = 1055 J).

While the K factor of PUR (0.11) may be used for comparative purposes commercially, the unit $W/m^2 \cdot K$ is preferred. Thus, PUR foam has a value of $0.023$ $W/m^2 \cdot K$ compared to fibrous glass and corkboard, with values of 0.035 and 0.039 $W/m^2 \cdot K$, respectively. A comparison of various properties of polyurethane, polystyrene, and polyvinyl chloride foams is given in Table 5.1.

Polyurethane can be applied as a spray-up foam in which the reactants are mixed at the gas nozzle. A comparable technique, in which the reactants are forced under high pressure into an impingement chamber, where they are intimately mixed at a lower pressure in the mold cavity, is called reaction injection molding (RIM). In the RIM process, reactants are injected toward each other at high speeds from opposite sides of the reaction chamber to ensure countercurrent (turbulent) molding at moderate temperatures (50 °C). This system, which may include reinforcements, is used to produce large automotive parts, such as bumpers, fenders, and

| Table 5.1  Properties of Typical Rigid Polymeric Foams | | | |
|---|---|---|---|
| **Property** | **PUR** | **PS** | **PVC** |
| Thermal conductivity, W/m² · K | 0.20 | 0.035 | 0.023 |
| Maximum service temperature, °C | 125 | 75 | 120 |
| Tensile strength, MPa | 1.2 | 0.7 | 1.2 |
| Flexible strength, MPa | 1.5 | 1.1 | 1.2 |
| Compressive strength, MPa | 30 | 15 | 35 |
| Dielectric constant | 1.1 | 1.02 | ... |
| Moisture vapor transmission, G/(m · s · GPa) | 35 | 30 | 15 |

side panels. In addition to PUR, the RIM process has been used to mold nylon 6 by the anionic polymerization of $\delta$-caprolactam in the presence of lithium (LiCl) or calcium chloride ($CaCl_2$). In 1989, 94,000 tons of PUR produced by RIM was used in the U.S., much of it by the automobile industry. These products consisted of hard segments produced by the reaction of methylene-4,4'-diphenyl diisocyanate (MDI) and extenders such as ethylene glycol and soft segments produced by the reaction of MDI and hydroxyl-terminated polyethers. The PUR products, which are also called integral skin foams, are characterized by a porous core and solid skin. These products have specific gravities ranging from 0.12 to 1.0.

Reaction injection molding, or the German equivalent RSG, is based on the development of self-skinning foam technology. Recent improvements in RIM technology have resulted in fast cycles that are nearly competitive with injection molding. New RIM machines have made possible the molding of extremely large parts (40 kg).

In addition to polyurethane RIM and structural RIM (SRIM), these processing techniques are now being used with dicyclopentadiene (Metlon), nylon 6 (Nyrim), epoxies, acrylamate/arylesters (Arimax), unsaturated polyester polyol-MDI reaction products (Xycon), modified polyisocyanurate (Arset), polyureas, polyurea/amides, and polyester/amides.

Chrysler has produced more than 20,000 SRIM truck beds, weighing 120 lb each.

The first generation of PUR, now called conventional RIM (CRIM), required long molding cycles and was based on the reaction of MDI, polyether polyols, and chain extenders (ethylene glycol). In the second-generation high-speed RIM systems, aromatic diamines (ethyltoluenediamine, ETDA) were used in place of some of the ethylene glycol, but a catalyst was essential for increasing the rate of urethane formation over that of polyurea. Internal release agents, such as solubilized zinc stearate, were used in third-generation RIM. Fourth-generation RIM's are polyurea elastomers produced by the reaction of amine-capped polyalkylene oxide and amine chain extenders with MDI-based isocyanates, in the absence of a catalyst. The polyurea/amide elastomers are produced by the reaction of ETDA with ketemines, which are the reaction product of aliphatic polyether and aliphatic ketones. Bayer is producing urethane/epoxy PUR (Desmocap) epoxies. Reinforced RIM (RRIM) has been produced by the addition of milled glass, fiberglass, or colloidal silica to PUR. The flexural modulus of RRIM is as much as 2.5 times greater than that of RIM.

ASTM has proposed five classifications for PUR coatings: uralkyd, moisture cure and unblocking cure (both of which are one-package systems), and catalyst cure and polyol cure (two-package systems). The moisture cure and the polyol cure systems are dependent on the reaction of terminal isocyanate groups with water and hydroxyl groups, respectively.

## Polystyrene

Rigid polystyrene (PS) foam is also an important commercial product. More than 1,300,000 tons of polystyrene are used annually in the U.S. As stated previously, preexpanded polystyrene beads are produced by the suspension polymerization of styrene in the presence of 5% pentane. Other EPS is produced by the injection of pentane or nitrogen into molten PS in an extruder prior to passing the EPS through a die to produce profiles, sheet, or board.

## Polyvinyl Chloride

Rigid PVC foams, produced at an annual rate of 23,000 tons in the U.S., have a higher specific gravity than PUR foams but a much lower specific gravity than solid PVC. Hence, considerable quantities of expandable PVC are extruded as a core in PVC pipe and sheet. This foam is also used as insulating board and sheet.

### Other Plastics

As stated previously, any solid polymer can be fabricated as a cellular product, including structural foams. Thus, acrylonitrile-butadiene-styrene (ABS), polycarbonate, polyphenylene oxide, polypropylene, and polysulfone, with a specific gravity of 0.80, are used as thermoplastic structural foams. Other typical structural thermoplastic foams have the following specific gravities: polyester, 1.10; cellulose acetate, 1.0; and polyethylene, 0.61. The annual volume of polyolefin foam is about 200,000 tons in the U.S., but should grow to more than 1 million tons by 1992.

Cellular ebonite, which was produced in the early 1920s, is one of the oldest rigid polymeric foams. Urea-formaldehyde (UF) foams have been used as "foamed-in-place" insulation in industrial and residential buildings. However, because of the possibility of release of small amounts of formaldehyde, this use has been drastically reduced in the U.S. However, phenol-formaldehyde, epoxy, polyimide, and nylon continue to be used as cellular polymers, and sprayed UF foam is used for daily covering of landfills.

## FLEXIBLE FOAMS

Almost 600,000 tons of flexible polyurethane is used annually in the U.S. for furniture, transportation, bedding, and carpet underlays. The chemistry is similar to that described for rigid PUR, except that a more flexible polyester or polyether with terminal hydroxyl groups is used.

Foamed flexible PVC can be produced by adding chemical blowing agents such as ABFA to PVC plastisols. These flexible products are used as textile coatings for clothing and furniture. Flexible foams are also produced from rubber latex and silicone elastomers. Closed-cell thermoplastic elastomer (TPE) foam is used for weatherstripping.

## TYPICAL APPLICATIONS

Open-cell flexible foams are used for cushioning in furniture, bedding, and shoe soles. Closed-cell low-density rigid foams are used for thermal insulation in refrigeration, transportation, and construction. Both foamed-in-place and loose-fill rigid foams are used for packaging. Foamed polystyrene is also used for food trays, egg cartons, and drinking cups.

Structural foams are being used in appliances, automobiles, construction, and furniture. Closed-cell rigid polymers are used for flotation applications, such as boats, docks, and buoys. Rigid cellular polymers have also been used in coaxial cables, radar domes, and sound-absorbing panels. Flexible foams have been used for gasketing, sealants, and closures. These and other applications of these lightweight materials will increase as designers become more familiar with the properties of these versatile products.

## REFERENCES

- J.K. Backus *et al.*, (ASTM classification), *Encyclopedia of Polymer Science and Engineering*, Vol. 13, John Wiley & Sons, New York, 1988
- A. Barker, (RIM), *Elastomers*, Vol. 122 (No. 1), 1990, p. 18
- J.A. Brydson, *Plastic Materials*, (RIM), Butterworths, London, 1989
- R. Burns, (RIM), *Plast. Technol.*, Vol. 34 (No. 11), 1988, p. 37
- R. Burns and J.K. Rogers, (Polyurea foam), *Plast. Technol.*, Vol. 34 (No. 10), 1988, p. 79
- D.G. Cogan, "Stones in a Glass House: CFC's and Ozone Depletion," Investors Responsibility Resource Center, Washington, DC, 1988
- W.J. Farrisey and K.W. Rausch, (Polyester/amide foam), *Elastomerics*, Vol. 120 (No. 7), 1988, p. 22
- T.H. Ferrigno, *Rigid Plastic Foams*, Reinhold, New York, 1967
- K.C. Frisch and J.H. Saunders, *Plastic Foams*, Part I and II, Marcel Dekker, New York, 1972, 1973
- M. Grayson, Ed., *Encyclopedia of Composite Materials and Components*, John Wiley & Sons, New York, 1983
- H.F. Herpe, in *Engineered Materials Handbook*, Vol. 2, *Engineering Plastics*, (Structural foams), J.N. Epal, J.M. Margolis, S. Newman, and R.B. Seymour, Ed., ASM International, Metals Park, OH, 1988
- E. Hunerberg, (Structural foams), *Modern Plastics*, Vol. 66 (No. 11), 1989, p. 278
- J.E. Kresta, Ed., *Reaction Injection Molding*, ACS Symposium Series, American Chemical Society, Washington, DC, 1985
- R. Leaversuch, (HCFC's), *Modern Plastics*, Vol. 65 (No. 13), 1988, p. 30

- F.F. Lindsay, (Polyurea/amide foam), *Modern Plastics*, Vol. 66 (No. 9), 1989, p. 137
- K.F. Lindsay, (CFC's), *Modern Plastics*, Vol. 65 (No. 12), 1988, p. 126
- K.F. Lindsay, (Geolite system), *Modern Plastics*, Vol. 66 (No. 12), 1989, p. 128
- A. Makhijani and A. Bickel, *Saving Our Skies*, Institute for Energy and Environmental Research, Washington, 1988
- R.J. Martino, Ed., *Modern Plastics Encyclopedia*, McGraw-Hill, New York, 1988
- R. McBrayer, Polyurethane Foams, *Modern Plastics*, Vol. 66 (No. 11), 1989, p. 279
- M.H. Naitove, (Water as a blowing agent), *Plast. Technol.*, Vol. 34 (No. 13), 1988, p. 59
- M.H. Naitove, (HCFC's), *Plast. Technol.*, Vol. 35 (No. 10), 1989, p. 161
- J. Oertel, Ed., *Polyurethane Handbook*, Hanser Publishers, New York, 1985
- O.E. Otteosted, Ed., (RRIM), *J. Appl. Polym. Sci.*, Vol. 34, 1987, p. 2575
- D.G. Schlotterbeck, G. Matzke, P. Horn, and H.V. Schmidt, (Polyurea/amide foam), *Plast. Eng.*, Vol. 45 (No. 1), 1989, p. 37
- G.R. Smoluk, (Soft blowing agent), *Modern Plastics*, Vol. 67 (No. 1), 1990, p. 78
- J.A. Snellor, (RIM), *Modern Plastics*, Vol. 65 (No. 3), 1988, p. 81
- J.A. Stuart, (SRIM), *Modern Plastics*, Vol. 65 (No. 1), 1988, p. 15
- R.W. Tess and G.W. Poehlein, Ed., *Applied Polymer Science*, American Chemical Society, Washington, DC, 1985
- P.A. Toensmeier, (Water as a blowing agent), *Modern Plastics*, Vol. 65 (No. 9), 1988, p. 16
- B.C. Wendle, *Structural Foams*, Marcel Dekker, New York, 1985
- G. Wood, *The ICI Polyurethanes Book*, John Wiley & Sons, New York, 1987

# CHAPTER 6
# Polymeric Blends

## INTRODUCTION

Blending of fibers, elastomers, and plastics has been practiced by the polymer industry for several decades. The textile designer is well aware of the advantage of blending hydrophobic fibers, such as linear aromatic polyesters (e.g., polyethylene terephthalate, or PET), with hydrophilic fibers, such as cellulose. Rubber compounders blended virgin natural rubber with reclaimed rubber for many years to reduce costs and, since 1940, have blended styrene-butadiene elastomer (SBR) with natural rubber in order to extend the supply of the latter.

When petroleum was inexpensive, there was not much incentive for blending general-purpose polymers. However, various polyolefins were blended to improve specific properties, and elastomers were blended with polystyrene to increase the impact resistance of this brittle polymer. These blends or alloys may be homogeneous, with a single glass transition temperature $(T_g)$, or heterogeneous, with dispersed microphases having different $T_g$'s.

The U.S. market for specialty polymer blends and alloys is valued at over $450 million and will approach $1000 million in the '90s. The development of polymer blends or alloys is a cost-effective way to fill gaps in the performance of existing polymers, of improving processibility, heat resistance, toughness, and flame retardancy, and of increasing sales without major capital expenditures or capacity expansion.

Polymer blends, which are mixtures of 20 or more polymers or copolymers, may be miscible or immiscible. In contrast, the term "polymer alloy" is used to describe mixtures of incompatible polymers to which are added compatibilizers.

## PRINCIPAL BLENDS

The leading common polymer blends, i.e., polypropylene/ethylene-propylene terpolymer, acrylonitrile-butadiene-styrene/polyvinyl chloride (ABS/PVC), modified nylon, and polyphenylene oxide/polystyrene (PPO/PS), now account for more than 80% of the total volume of blends. However, because of increasing demands for these products, these "Big Four" blends will account for less than 70% in future years. PPO/PS blends, which are sold by General Electric under the trade name of Noryl, account for 43% of all engineering resin blends, nylon/elastomers for 10%, polycarbonate/high-density polyethylene (PC/HDPE) for 8.2%, and PC/ABS for 3.5%. The total annual dollar volume of these blends is $350 million. The rubber-modified blend of PC/PET is called Xenoy by G.E. Other commercial blends include PC/nylon 612, silicone/polyether-imide (PEI) (Ektar), nylon 66/ABS (Triax), PPO/PET (Genox), polycarbonate/polyurethane (PC/PUR) (Makrolon), ABS/PVC (Pellethan), polyphenylene ether (PPE)/nylon (Ultraryl), PPE/ASA (Terblet), nylon 66/ABS (Elemid), PVC/ABS (Cyrovin), PS/ABS (Mindel), PC/ABS (Pulse), and nylon 66/HDPE (Selar).

Currently, the transportation industry consumes 40% of all polymer blends, the business machine industry about 25%, and the appliance industry 11% of the total volume. These blends or alloys can be produced by melt mixing of two or more polymers in an intensive blender.

## COMPATIBILIZATION

In some instances, such as incompatible blends of PVC/polyethylene (PE), it is necessary to add a compatibilizing agent. In this case, chlorinated polyethylene (CPE) may be used as the compatibilizing agent.

In addition to miscible one-phase systems, such as PPE/PS, which have single glass transition temperatures $(T_g)$, there are numerous two-phase systems. If the submicroscopic regions are amorphous and larger than 5 nm, there will be two dampening peaks as well as two glass transition temperatures when the log of shear modulus of a two-phase system is plotted against temperature. The width of the dampening peak is related to the incompatibility of the two polymeric components of the blend. In general, polymers having solubility parameters differing by at least 2 Hildebrand units will be incompatible. Polyblends may be thermodynamically unstable, but any separation of components is slow because of the low diffusion coefficient.

According to J. Barlo, the best-known compatibilizing agents are polycaprolactone (PCL) and polybutylene adipate (PBA). The addition of one part of PBA compatibilizer is sufficient for blends of styrene-acrylonitrile/polycarbonate (SAN/PC), and a relatively high loading of PCL will miscibilize phenoxypolycarbonate resins. An acid-modified polyolefin has been used as the compatibilization agent for polyethylene and nylon 66. The law of additives may be applied to homogeneous blends whose properties are related to the relative amounts of the compatible polymers present.

Most compatibilizers are products that have been used previously for other applications, such as those used for tie layers in coextrusions. In many systems, a copolymer based on maleic anhydride, which acts as a "hook," is used as the compatibilizer. Maleic anhydride-modified thermoplastic elastomers (TPE) are used as compatibilizers for polypropylene (PP)/nylon blends. Maleic anhydride-modified olefinic copolymers are also used for PP/nylon blends and for other blends such as polyolefin/ethylene-vinylalcohol copolymer (EVOH), polystyrene/EVOH, and polyethylene/polyester. A terpolymer of ethylene-methyl/acrylate and maleic anhydride has been blended with nylon 66 to improve impact resistance.

Styrene maleic anhydride (SMA) copolymers, including copolymers such as SMA/acrylonitrile terpolymers (Cadon), are used to compatibilize PC/nylon blends. Acrylic acid modified PP is used to compatibilize PET and polyolefins, and PCL is used as a compatibilizer for PVC/PS and PC/SAN. Styrene butadiene block copolymers are used as compatibilizers for PS/polyolefin blends. Block copolymers of PCL and PC are used to compatibilize blends of PC/SAN and PVC/PS.

Imidized acrylics, which have previously been used as impact modifiers and coupling agents, are now being used as compatibilizers for PC/nylon. Phenoxy resins have been used as compatibilizers for blends of PC with ABS and PC/SMA.

Silanes have been used to compatibilize PPE/nylon and polybutylene terephthalate/epoxy resin (PBT/EP). Chlorinated polyethylene, which has been used as a compatibilizer with PS/HDPE and PVC/HDPE, is also used as a compatibilizer for mixed scrap plastics.

HDPE treated with an oxidizing gas to produce active sites on its surface has been blended with nylon, polyurethane, and epoxy resins. Di-

| Table 6.1 Properties of Compatibilized Nylon 6/PBT Blends | | |
|---|---|---|
| **Property** | **Nylon 6/PBT** | **Fiberglass-reinforced nylon 6/PBT** |
| Flexural strength, MPa | 90 | 165 |
| % Elongation | 19 | 2 |
| Flexural modulus, GPa | 2750 | 7950 |
| Heat-deflection temperature, °C | 165 | 215 |
| Impact resistance, J/m | 3925 | 14,585 |
| Note: J/m × 0.0187 = ft · lb/in. | | |

methylaminourethane (DMAU) has been used to increase the impact resistance of polymeric blends.

The properties of a 50/50 blend of nylon 6/PBT with an SMA/glycidyl methacrylate terpolymer compatibilizer are shown in Table 6.1.

## PPO/PS BLENDS

A. Hay of G.E. invented poly(2,6-dimethylphenyl)1,4-ether (polypropylene oxide, or PPO), which he produced by a unique copper ion-catalyzed oxidative coupling reaction of 2,6-dimethylphenol. Because of linear bonded aromatic ring structures, these amorphous polymers are stiff, have high melting points, and are difficult to process.

However, the processibility of PPO was improved by blending it with polystyrene (Noryl). Prior to its acquisition by G.E., Borg-Warner produced a copolymer of 2,6-dimethylphenol and 2,3,6-trimethylphenol, which was also blended with polystyrene to achieve good processibility. These PPO blends are sold by G.E. under the trade name of Prevex. The addition of PS decreases melt viscosity and heat-deflection temperatures of the homopolymer and the copolymer.

Asahi has been supplying an engineering resin to Dow Chemical Company under the trade name of Xyron. Asahi has patented this polymer in Japan; annual production is 70 million tons. Litigation between G.E. and Dow has been settled, and Dow still continues to supply a modified polymer that does not infringe on the G.E. patent. Another Japanese firm is also producing similar engineering resins. More than 70,000 tons of PPO are produced annually in the U.S.

Sulfonated polyphenylene oxide is being used as a thin fiber composite membrane for the purification of waste water in the Alberta tar sands. Blends of these engineering resins are also being used for television cabinets, computers, automotive panels, wheel covers, automobile spoilers, pump impellers, radomes, and many types of lighting devices.

PPO's are softened by chlorinated solvents and stress crack in the presence of ketones, esters, or aromatic hydrocarbons. The crystallization behavior of PPO/polymethyl methacrylate (PMMA) blends has been evaluated by the use of spectroscopic and calorimetric techniques. The dielectric constant of Noryl is essentially constant from 60 to $10^6$ cps and at temperatures below the heat-deflection temperature (100 °C). These blends have excellent resistance to cold flow under static and cyclic loadings. The addition of impact modifiers increases the impact resistance of PPO/PS blends over a wide range of temperatures. PPO/PS blends reinforced with 10% carbon fiber have high flexural moduli (4 GPa), and blends reinforced with 30% fiberglass also have high flexural moduli (7 GPa). Mold shrinkage is also reduced by the addition of these reinforcements. The thermal and physical properties of PPO/PS are shown in Table 6.2.

## RUBBER/PS (HIPS)

The existence of separate rubber and PS beta-dampening peaks and microscopy of natural rubber/polystyrene blends have demonstrated the existence of two phases and that rubber is dispersed in the PS matrix. The rubber phase in this two-phase blend is dispersed as fine particles that are stretched when stress is applied. These rubber particles tend to trigger the beta mechanism responsible for the crazing of the polymer blend. However, the particle size remains unchanged during processing, and this is essential for good performance in a heterophase polymer blend.

The extent of crazing in these two-phase systems is related to the number of particles with sizes greater than a threshold size. The crazing is related to impact strength, ductility, and the extent of stress whitening resulting from craze cracks in these polymeric composites. It is now recognized that energy absorption in an impact test is dependent on the

Table 6.2  Thermal and Physical Properties of PPO/PS

| Property | PPO/PS modified | 30% glass-filled PPO/PS | 30% graphite-filled PPO/PS |
|---|---|---|---|
| Glass transition temperature ($T_g$), °C | 125 | 120 | 115 |
| Heat deflection temperature at 1.82 MPa, °C | 100 | 145 | 130 |
| Maximum resistance to continuous heat, °C | 80 | 130 | 105 |
| Coefficient of linear expansion, cm/cm/°C $\times 10^{-5}$ | 5.0 | 2.0 | 1.0 |
| Tensile strength, MPa | 55 | 120 | 128 |
| % Elongation | 50 | 4 | 2.5 |
| Flexural strength, MPa | 89 | 144 | 138 |
| Compressive strength, MPa | 96 | 123 | 130 |
| Notched Izod impact strength, J/m | 270 | 107 | 72 |
| Hardness, Rockwell | R115 | R115 | R111 |
| Specific gravity | 1.1 | 1.27 | 1.25 |

interfacial separation of the rubber particles from the PS and crazing. The relaxation strength ($S_m$) is also related to impact strength and is equal to the difference in shear modulus (G) at different temperatures divided by the shear modulus at the higher temperature ($T_2$).

Buchdahl and Nielsen have proposed the following generalizations for high-impact polystyrenes (HIPS):

- The $T_g$ of the rubber component must be less than 50 °C.
- Occlusion of PS extends the dispersed rubber phase.
- A variation in the degree of crosslinking of the rubber may optimize the ultimate properties of HIPS.
- The size of the rubber particles should be in the range of 1 mm or less than 5 mm.

Table 6.3  Properties of Specialty HIPS

| Property | Ignition HIPS | Very high impact-resistant HIPS | High-gloss HIPS | Fiberglass-filled (10%) HIPS |
|---|---|---|---|---|
| Tensile strength, MPa | 27 | 19 | 23 | 40 |
| Tensile modulus, GPa | 1.8 | 1.7 | 1.7 | 2.5 |
| % Elongation | 40 | 50 | 35 | 4 |
| Flexural strength, MPa | 40 | 33 | 43 | 58 |
| Flexural modulus, GPa | 2 | 1.9 | 1.9 | 3.2 |
| Izod impact strength, J/m | 107 | 214 | 133 | 133 |
| Heat-deflection temperature, °C | 87 | 87 | 85 | 92 |
| Specific gravity | 1.15 | 1.0 | 1.0 | 1.13 |
| Flammability | V.0 | ... | ... | ... |
| Gardner gloss | ... | ... | 85 | ... |

- The interface between the two phases is less than optimum when the number of occlusions is low, as is the case with mechanical blending of PS and rubber.
- The impact resistance of the composite will vary with areas unless isotropic test samples are used.
- The coefficient of thermal expansion is a function of the rubber content.

More than 150,000 tons of HIPS is used annually in the U.S. The properties of specialty HIPS and a typical HIPS are listed in Tables 6.3 and 6.4.

## BLENDS OF STYRENE COPOLYMERS

Styrene copolymers, such as styrene-maleic anhydride (SMA), styrene-acrylonitrile (SAN), acrylonitrile-styrene-acrylic esters (ASA), and acrylonitrile-butadiene-styrene (ABS), as well as blends of these copolymers,

| Table 6.4  Properties of a Typical HIPS | |
|---|---|
| **Property** | |
| Glass transition temperature ($T_g$), °C | 90 |
| Heat-deflection temperature at 1.820 MPa, °C | 90 |
| Maximum resistance to continuous heat, °C | 70 |
| Coefficient of linear expansion, cm/cm/°C $\times$ $10^{-5}$ | 8.0 |
| Tensile strength, MPa | 41 |
| % Elongation | 3 |
| Flexural strength, MPa | 50 |
| Compressive strength, MPa | 45 |
| Hardness, Rockwell | R65 |
| Specific gravity | 104 |
| Impact strength, J/m | 80 |

are commercially available. ABS, which is the most widely used copoly-mer blend, is produced at an annual rate of about 500,000 tons in the U.S. Annual worldwide consumption of ABS is more than 1.3 million tons.

## ABS Copolymers

ABS is produced by G.E., Dow, and Monsanto under the trade names of Cycolac, Magnum, Rovel, and Elite. The original ABS, which was composed of a synthetic rubber such as acrylonitrile-butadiene rubber (NBR) and PS, has been modified by numerous modifications, all labeled by the same acronym ABS. The theories for ductility and impact strength of ABS are similar to those discussed for HIPS previously in this chapter. The properties of typical ABS plastics are listed in Table 6.5.

ABS is also subject to stress cracking when exposed to organic liquids. There is little effect when ABS is exposed to alkalies or organic acids,

Table 6.5  Properties of Typical ABS Plastics

| Property | Extrusion grade | 20% glass-reinforced |
|---|---|---|
| Glass transition temperature ($T_g$), °C | 115 | 125 |
| Heat-deflection temperature at 1.82 MPa, °C | 90 | 100 |
| Maximum resistance to continuous heat, °C | 80 | 90 |
| Coefficient of linear expansion, cm/cm/°C $\times 10^{-5}$ | 9.5 | 2.0 |
| Tensile strength, MPa | 34 | 76 |
| % Elongation | 60 | 5 |
| Flexural strength, MPa | 63 | 103 |
| Compressive strength, MPa | 48 | 96 |
| Notched Izod impact strength, J/m | 320 | 53 |
| Hardness, Rockwell | R60 | M85 |
| Specific gravity | 1.03 | 1.2 |
| Dielectric constant | 3.0 | 3.0 |

but resistance to stress cracking is poor in solvents, such as toluene and ethyl acetate.

Blends of ABS/nylon (Elemid), ABS/polycarbonate (Bayblend), ABS/PVC (Kralastic), ABS/polysulfone (PSU) (Mindel), and ABS/SMA (Cadon) are commercially available. The alloys of ABS and high-performance polymers are more readily processed than the engineering polymers. ABS/PVC alloys have outstanding impact strength, and ABS/PSU alloys are tough plastics. Mindel has a heat-deflection temperature of 150 °C. Properties of several commercial ABS blends are listed in Table 6.6.

ABS terpolymers compete with PVC for the extruded pipe market, but blends of these two resins are also widely used. PVC improves flame retardancy, chemical resistance, and rigidity, and ABS provides improved

| Table 6.6 Properties of Commercial ABS Blends | | | | |
|---|---|---|---|---|
| Property | ABS/PVC | ABS/PC | ABS/SMA | ABS/PA |
| Tensile strength, MPa | 40 | 52 | 36 | 36 |
| Tensile modulus, GPa | 2.5 | 2.5 | 2.2 | 1.3 |
| Flexural strength, MPa | 65 | 85 | 60 | ... |
| Flexural modulus, GPa | 2.4 | 2.3 | 2.2 | 1.0 |
| Izod impact resistance, J/m | 430 | 450 | 160 | 960 |
| Hardness, Rockwell | R100 | R115 | R100 | R75 |
| Specific gravity | 1.05 | 1.03 | 1.03 | 1.06 |
| Heat-deflection temperature, °C | 77 | 113 | ... | 92 |
| Coefficient of expansion, $10^{-5}/°C$ | 8.3 | 6.7 | 9.0 | 10 |
| UL temperature index, °C | 65 | 95 | 50 | 60 |

processibility and impact resistance. G.E. and Monsanto are the leading American suppliers of these PVC/ABS blends. G.E., the world's largest supplier of engineering resins, is supplying a blend of PVC and ASA, called Geloy (see Table 6.7). The ASA production facilities were acquired by G.E. from Stauffer. The ABS facilities were acquired from Borg-Warner.

## SAN Copolymers

In addition to being blended with NBR for some types of ABS, SAN is also blended with ethylene-propylene elastomers (EPDM, Rovel). However, the ASA copolymer is used in many other blends, such as with PC to produce terblends, with PVC, and with PMMA.

| Table 6.7 Properties of a Typical ASA/PVC Blend (Geloy) | |
|---|---|
| **Property** | |
| Heat-deflection temperature at 1.82 MPa, °C | 77 |
| Maximum resistance to continuous heat, °C | 70 |
| Coefficient of linear expansion, cm/cm/°C × 10$^{-5}$ | 8.8 |
| Tensile strength, MPa | 26 |
| % Elongation | 25 |
| Flexural strength, MPa | 45 |
| Notched Izod impact strength, J/m | 1088 |
| Hardness, Rockwell | R96 |
| Specific gravity | 1.4 |

## SMA Copolymers

The original terpolymer of styrene, maleic anhydride, and acrylonitrile (Cadon) has been modified by blending with elastomers and other polymers. SMA has also been blended with PC (Arloy), PVC, and HIPS (Dylark). As described for ABS, there are many different formulations for Cadon, Arloy, and Dylark.

In general, SMA increases the temperature resistance of blends. The addition of SMA to PC increases the heat-deflection temperature by 10 to 15 °C. The properties of a typical SMA/PC alloy and fiberglass-filled SMA/ABS blends are shown in Tables 6.8 and 6.9.

Toughened nylon 6 has been produced at the expense of tensile strength by the solution polymerization of caprolactam in the presence of ethylene-propylene elastomers (EPM) as well as by melt blending of the two polymers. These blends require good adhesion between the nylon and EPM. The toughness is inversely related to the size of the rubber particles. Nylon 66 has been toughened by the addition of maleic anhydride grafted polyethylene. Best performance is obtained by the blending of 0.35-mm-diam particles of 30% grafted PE.

| Table 6.8  Properties of a Typical SMA/PC Alloy | |
|---|---|
| **Property** | |
| Tensile strength, MPa | 48 |
| % Elongation | 85 |
| Flexural strength, MPa | 93 |
| Flexural modulus, GPa | 2.4 |
| Izod impact resistance, J/m | 800 |
| Heat-deflection temperature, °C | 113 |
| Specific gravity | 1.13 |

| Table 6.9  Properties of Typical Fiberglass-Filled SMA/ABS Blends | | |
|---|---|---|
| **Property** | **10% fiberglass** | **20% fiberglass** |
| Tensile strength, MPa | 70 | 90 |
| Flexural strength, MPa | 115 | 135 |
| Flexural modulus, GPa | 0.5 | 0.6 |
| Heat-deflection temperature, °C | 105 | 115 |
| Coefficient of expansion, $10^5/°C$ | 4.3 | 4.3 |

## NYLON BLENDS

The impact strength of both nylon 66 and nylon 6 has been improved and the tensile strength and flexural modulus have been decreased by blending with EPDM (Zytel), other elastomers (Ultramid), ethylene co-polymers (Capron, Nycoa, Grilon), and polytetrafluoroethylene (PTFE) (Nylafil, Electrafil). The immiscible mixture of nylon 66 and HDPE (Selar) has been used as a barrier coating. ASA/nylon blends have higher heat deflection values than ABS/nylon blends. Blends with fillers have higher heat deflection values and are stiffer than unfilled blends.

| Table 6.10  Properties of a Typical PC/PBT Blend | |
|---|---|
| **Property** | |
| Heat-deflection temperature at 1.82 MPa, °C | 90 |
| Tensile strength, MPa | 55 |
| % Elongation | 120 |
| Flexural strength, MPa | 86 |
| Notched Izod impact strength, J/m | 120 |
| Specific gravity | 1.23 |

## POLYCARBONATE BLENDS

In addition to the ABS/PC blends, polycarbonate has been blended with polyesters, SMA copolymers (Arloy), and polyethylene and polyurethane (Texin). The PC/polyolefin blends are less notch sensitive than PC. PC/PUR blends also have high impact resistance at low temperatures. The PC/SMA blends are also characterized by good low-temperature impact resistance, heat resistance, and a relatively high modulus.

PET alloys are transparent and tough and are characterized by good dimensional stability at varying degrees of humidity and by good resistance to gasoline and hydraulic fluids. Blends of PC/PBT (Xenoy), which may also contain some elastomeric polymers, have properties similar to PC/SMA and are resistant to aliphatic solvents such as gasoline (see Table 6.10). Hollow graphite-reinforced PC/PBT blends, filled with polyurethane foam, are being used for baseball bats.

## POLYESTER BLENDS

In addition to PC/PBT and PC/PET blends, the following alloys are commercially available: PBT/PET (Celanex), PBT/elastomer (Gafite), Pocan, Duraloy, Ultradur, PET/elastomer (Rynite), PET/PMMA (Ropel), and PEI/PSU.

The addition of PET to PBT lowers costs and increases the rate of crystallization. PET/elastomer and PBT/elastomer blends have high impact strengths and improved toughness and modulus. Fiberglass-filled

| Table 6.11 Properties of Typical HDPE Blends | | |
|---|---|---|
| **Property** | **HDPE** | **HDPE/nylon + compatibilizer (70:20:10)** |
| Tensile strength, MPa | 27 | 34 |
| % Elongation | 300+ | 300 |

PET/PSU alloys have low mold shrinkage, good heat resistance, high modulus, and good dimensional stability.

## OTHER COMMERCIAL BLENDS

Blends of acetal/elastomer (Duraloy, Ultraform) have improved toughness and fatigue resistance. Blends of acetal/PTFE (Formaldafil) are characterized by good lubricity and wear resistance. Blends of PP/EPDM (Santoprene, Prolastic, Feroflex), which are called thermoplastic polyolefins (TPO), have improved toughness, puncture resistance, and stress crack resistance.

Blends of HDPE with LLDPE have improved tear resistance, puncture resistance, and heat sealability. Blends of HDPE and ionomers (Surlyn) are characterized by good processibility, good heat resistance, and excellent impact strength.

Styrene-modified PP has improved mechanical properties, similar to the additivity of both polymers. This blend aids the compatibilization of a blend of PP and HIPS. As shown in Table 6.11, anhydride or acid-modified polyolefins and dispersions of polyolefins and nylon 66 have improved tensile-strength properties.

PTFE and other polyfluorocarbons are being blended with fluoroelastomers (Fluocel, Viton) to lower costs and improve properties of these heat- and corrosion-resistant products. Prior to World War II, recycled rubber was blended with natural rubber to improve processibility and reduce costs. A blend of epoxidized rubber and natural rubber has been proposed as an elastomer for better performance under wet or icy conditions.

## PVC BLENDS

The impact resistance of rigid PVC is improved by blending with SMA (Elemid), chlorinated/polyethylene (Hoslalite), and ethylene-vinyl acetate copolymer (Sumagraft). Varying degrees of flexibility are obtained by blending PVC with NBR (Vynite). PVC/PUR alloy (Vythene) is characterized by good light stability and low-temperature flexibility.

## CHARACTERIZATION OF
## PLASTIC BLENDS

The solid-state phase behavior of strongly interacting polymer blends has been correlated by high-resolution $^{13}C$ NMR and FTIR spectroscopy. Differential scanning calorimetry (DSC) has been used to monitor melting point transition and to construct temperature composite projections for binary phase diagnosis.

## REFERENCES

- D. Aycock, (Reinforced PPO/blends), *Modern Plastics*, Vol. 66 (No. 11), 1989, p. 90
- P. Bruins, *Polyblends and Composites*, Interscience Publishers, New York, 1970
- C. Cheng and L.A. Belfiore, (Solid state behavior), *Polymer News*, Vol. 15, 1990, p. 39
- J.L. Cooper and G.M. Estes, Ed., *Multiphase Polymers*, Advances in Chemistry Series 176, American Chemical Society, Washington, DC, 1979
- M.C. Gabriele, (Compatibilization), *Plast. Technol.*, Vol. 36 (No. 5), 1990, p. 25
- D. Klemper and K.C. Frisch, *Polymer Alloys*, Vol. I and II, Plenum Press, New York, 1977, 1979
- M.A. Kohudic, Ed. *Advances in Polymer Blends and Alloys Technology*, Technonic Publishing, Lancaster, PA, 1988
- J.I. Kroschwitz, Ed., *Encyclopedia of Polymer Science and Engineering*, Vol. I, Wiley-Interscience, New York, 1985
- J.A. Manson and H. Sperling, *Polymer Blends and Composites*, Plenum Press, New York, 1976
- R.J. Martino, Ed., *Modern Plastics Encyclopedia*, McGraw-Hill, New York, 1986

- M.H. Naitove, (Compatibilization), *Plast. Technol.*, Vol. 35 (No. 2), 1989, p. 67
- O. Olabisi, L.M. Robeson, and M.T. Shaw, *Polymer/Polymer Miscibility*, Academic Press, New York, 1979
- D.R. Paul and S. Newman, Ed., *Polymer Blends*, Academic Press, New York, 1978
- A.J. Platzer, Ed., *Copolymers, Polyblends and Composites*, Advances in Chemistry Series 142, American Chemical Society, Washington, DC, 1975
- R.B. Seymour, *Plastics vs. Corrosives*, Wiley-Interscience, New York, 1982
- R.B. Seymour and G.S. Kirshenbaum, Ed., *High Performance Polymers: Their Origin and Development*, Elsevier Science, New York, 1986
- J. Wigotsky, (Blends), *Plast. Eng.*, Vol. 44 (No. 11), 1988, p. 25

# CHAPTER 7

# Thermosetting Plastics

## INTRODUCTION

Prior to the 1940s, the production of thermosetting plastics (thermosets) was greater than the production of thermoplastics. However, because of lower production costs and ease of fabrication via injection molding and extrusion, which are less labor intensive than compression molding, production of thermoplastics now exceeds that of thermosets. Most of the first man-made composites consisted of fillers and thermosetting (cross-linking) polymers, such as filled ebonite, filled vulcanized rubber, and wood flour-filled phenolic resins (Bakelite). Nevertheless, while cellulose nitrate was intractable, it was not a thermoset. Celluloid was a mixture of camphor and the thermoplastic cellulose nitrate. Yet, most of the early plastic technologists believed that composites of resin and filler could be produced only from thermosets. Thus, urea formaldehyde (Beetle, Plaskon), melamine-formaldehyde (Melmac), and unsaturated polyesters were used as reinforced composites, but the pioneer thermoplastics, such as acrylics, polystyrene, and polyvinyl chloride (PVC), were generally used without fillers or reinforcements.

The thermosets will be discussed in alphabetical order in this chapter.

## ALKYDS

Since the generic name is derived from abbreviated names of the reactants *al*cohols and a*cids*, all polyesters may be classified as alkyds. However, this name is generally used to describe condensation products of polyhydric alcohols and polybasic acids, modified by monobasic fatty acids. Since the objective is to produce a linear polymer with the highest possible molecular weight, with low acid numbers, and without gelation, Patton has modified the Carothers' equation to include an "alkyd content" (K), which is equal to the total number of carboxyl- and hydroxyl-containing molecules in the reactants ($m_o$) divided by the total number of acid equivalents present ($e_A$). This equation, in which K = 1, and the Carothers' equation, in which DP is degree of polymerization, p is extent of reaction, f is functionality, and c is concentration, are:

$$K = \frac{m_o}{e_A} = 1 \qquad \text{(Patton equation)}$$

$$DP = \frac{c}{1 - (pf/2)} \qquad \text{(Carothers' equation)}$$

Patton showed that alkyds produced from phthalic anhydride have low K values (<1.02) and that those produced from the para dicarboxylic acid (terephthalic acid) have higher K values (>1.05). Low K values indicate the probability of premature gelation before completion of the esterification.

According to the Carothers' equation, a DP of infinity (crosslinked polymer) is obtained with a 95.6% yield (p = 0.956) in the presence of 0.1 mole of trifunctional glycerol, and incipient gelation (crosslinked polymerization) is unavoidable even at 87.5% yield when 0.2 moles of glycerol is present.

Whether one uses the Carothers' or Patton equation, it will be obvious that the degree of crosslinking can be controlled by functionality. It can also be reduced by the addition of monofunctional reactants. It is important to note that these concepts will hold for all polymerization reactions in which one of the reactants has a functionality greater than 2. It should be noted that oleic acid $(CH_3)(CH_2)_7CH=CH(CH_2)_7(COOH)$ has a functionality of 1, that the functionality of ethylene glycol $((H_2COH)_2)$ is 2, and that a mixture of 0.8 moles of ethylene glycol and 0.2 moles of glycerol $(CH_2OHCH(OH)CH_2OH)$ is 2.2.

The molecular weight of linear polyesters can be controlled by the relative amounts of the difunctional reactants. The general equation for a linear condensation (step reaction) polymerization is:

$$nXRX + nYR'Y \rightarrow nXY + X(R\ R')nY$$

The general equation for a theoretical condensation polymerization is:

$$nXRX + nYR''Y \rightarrow nXY + (XR\overset{Y}{R''}Y)_n \rightarrow \text{a network polymer}$$

The properties of a typical glass-filled alkyd are shown in Table 7.1.

## ALLYLIC POLYMERS
## (DIALLYL POLYMERS)

Because of the lack of a strong activating substituent and the presence of a third carbon atom with more reactive hydrogen atoms, monomers with one allylic (2-propenyl, $H_2C=CHCH_2X$) group are not readily polymerizable by classic chain-reaction polymerization techniques. However, some allylic monomers may be polymerized in the presence of Ziegler catalysts. The name "allyl" is derived from the Latin *allium* (garlic), which contains allyl isothiocyanate ($C_3H_5NCS$). Diallyl compounds are readily polymerized by free radical initiators. The first commercial allyl polymer, CR-39, was produced by the polymerization of diethylene glycol *bis* (chloroformate) (DADC), $O=C(CH_2CH_2OCOOCH_2CH=CH_2)_2$.

This polymerization, like all chain-reaction polymerizations, proceeds through initiation, propagation, and termination steps, as shown in the following equations in which the initiator is a free radical (R•) and M represents the monomer:

$$R\bullet + M \rightarrow RM\bullet \qquad \text{(Initiator)}$$

$$RM\bullet + nM \rightarrow RM_nM\bullet \qquad \text{(Propagation)}$$

$$2RM_nM\bullet \rightarrow RM_nMMM_nR \qquad \text{(Termination by coupling)}$$

The properties of a typical fiberglass-filled alkyd are shown in Table 7.2.

Diallyl phthalate (DAP) is also readily polymerized by free radical initiators, such as benzoyl peroxide (BPO). The isophthalate (meta isomer) polymerizes more rapidly than other DAP isomers. Polymers of DAP are used for molding electrical devices. The properties of typical filled and unfilled DAP are shown in Table 7.3.

Diallyl maleate (DAM), diallyl fumarate (DAF), and triallyl cyanurate (TAC) may also be polymerized to produce crosslinked polymers, which are also called thermosets. DAM and DAF have been used as comono-

| Table 7.1  Properties of a Typical Glass-Filled Alkyd | |
| --- | --- |
| **Property** | |
| Heat-deflection temperature at 1.82 MPa, °C | 200 |
| Maximum resistance to continuous heat, °C | 200 |
| Coefficient of linear expansion, cm/cm/°C $\times 10^{-5}$ | 2.0 |
| Tensile strength, MPa | 41 |
| Flexural strength, MPa | 103 |
| Notched Izod impact strength, J/m | 15 |
| Hardness, Rockwell | E80 |
| Specific gravity | 2.1 |

| Table 7.2  Properties of a Typical CR-39 Polymer | |
| --- | --- |
| **Property** | |
| Glass transition temperature ($T_g$), °C | 85 |
| Heat-deflection temperature at 1.82 MPa, °C | 60 |
| Maximum resistance to continuous heat, °C | 100 |
| Coefficient of linear expansion, cm/cm/°C $\times 10^{-5}$ | 8.1 |
| Tensile strength, MPa | 31 |
| Flexural strength, MPa | 55 |
| Notched Izod impact strength, J/m | 15 |
| Hardness, Rockwell | M95 |
| Specific gravity | 1.31 |

Table 7.3  Properties of DAP

| Property | DAP | Fiberglass-filled | Mineral-filled |
|---|---|---|---|
| Heat-deflection temperature at 1.82 MPa, °C | 155 | 225 | 200 |
| Maximum resistance to continuous heat, °C | 100 | 210 | 200 |
| Coefficient of linear expansion, cm/cm/°C $\times 10^{-5}$ | ... | 3 | 3.5 |
| Tensile strength, MPa | 27.6 | 50 | 45 |
| % Elongation | 4.6 | 4 | 4 |
| Flexural strength, MPa | 62.0 | 90 | 65 |
| Compressive strength, MPa | 150 | 205 | 170 |
| Notched Izod impact strength, J/m | 14 | 50 | 16 |
| Hardness, Rockwell | E115 | E84 | E61 |
| Specific gravity | 11.3 | 1.80 | 1.75 |

mers with unsaturated polyesters. TAC is used in small amounts to cross-link other vinyl monomers.

## AMINO PLASTICS

Urea formaldehyde (UF) resins reinforced with alpha-cellulose were introduced in the 1920s by E.C. Rossiter, under the trade name of Beetle. Melamine formaldehyde (MF) resins, which are more temperature resistant than UF, were introduced in the 1930s. Both resins are dependent on the introduction of hydroxymethyl (methylol) groups formed by a reaction with formaldehyde (HCHO), as shown by the following equation:

$$RNH_2 + HCHO \rightarrow RNHCH_2OH$$

In subsequent steps, these methylol groups ($-CH_2OH$) are condensed with the amino groups ($NH_2$) to produce water and a product with a methylene bridge, as shown by the following equation:

| Table 7.4  Properties of Typical Amino Plastics | | |
|---|---|---|
| Property | Cellulose-filled UF | Fiberglass-filled MF |
| Heat-deflection temperature at 1.82 MPa, °C | 175 | 200 |
| Maximum resistance to continuous heat, °C | 100 | 175 |
| Coefficient of linear expansion, cm/cm/°C $\times 10^{-5}$ | 4.0 | 1.6 |
| Tensile strength, MPa | 65 | 50 |
| % Elongation | 0.7 | 0.7 |
| Flexural strength, MPa | 90 | 130 |
| Compressive strength, MPa | 275 | 310 |
| Notched Izod impact strength, J/m | 16 | 100 |
| Hardness, Rockwell | M120 | M120 |
| Specific gravity | 1.50 | 1.9 |

$$RNHCH_2OH + H_2NR \rightarrow RNHCH_2NHR + H_2O$$

Urea ($H_2NCONH_2$) has four reactive hydrogen atoms, two of which are capable of producing linear polymers, and two others, by subsequent reaction, capable of producing crosslinked polymers. Melamine ($C_3N_3(NH_2)_3$) is a cyclic triazine with six hydrogen atoms which undergo reactions with formaldehyde, comparable to those outlined for urea. Both of these amino resins may be molded to produce light-colored articles with good electrical properties and fair corrosion resistance. The properties of typical cellulose and fiberglass-filled amino resins are shown in Table 7.4. In 1989, 621,000 tons of amino resins were produced in the U.S.

## CYANATE ESTERS

Commercial high-performance polymers are produced by the cyclo-trimerization of di- or polyfunctional aryl cyanate esters to s-triazines.

| Table 7.5  Properties of Typical Cyanate Ester Plastics | | |
|---|---|---|
| Property | Cyanate ester | Toughened bismaleimide |
| Tensile strength, MPa | 83 | 55 |
| Tensile modulus, GPa | 3.2 | 3.5 |
| Glass transition temperature ($T_g$), °C | 289 | ... |
| Coefficient of expansion, $10^{-5}$/°C | 6.4 | 6.0 |
| Heat-deflection temperature, °C | 245 | 275 |
| Dielectric constant | 43 | 3.5 |

These aryl cyanate esters can be produced as liquid or solid crosslinkable prepolymers. Aryl cyanate esters, such as bisphenol A dicyanate, can be processed like epoxy resins. These polymers, which have a heat-deflection temperature of 250 °C, can be reinforced with carbon fibers and used as advanced structural composites. The properties of typical cyanate ester plastics are shown in Table 7.5.

## EPOXY RESINS

Crosslinkable epoxy (EP) resins, based on the condensation of bisphenol A ($p,p'$-isopropylidene diphenol), $(HOC_6H_4)_2C_3H_6$ (BPA) and epichlorohydrin ($ClCH_2CHOCH_2$), were produced by DeTrey Freres in 1936. EP resins may be cured at ordinary temperatures by Lewis bases, such as amines, which react with the terminal oxirane (epoxy) groups. Primary amines react twice as fast as secondary amines, and this rate is accelerated in the presence of hydroxyl compounds. EP resins may also be cured at elevated temperatures (200+ °C) by the addition of cyclic anhydrides, which react with the hydroxyl pendant groups.

EP resins may also be crosslinked by the addition of amino or phenolic resins. More than 10,000 tons of polyamides and 8000 tons of amino and phenolic resins are used annually as EP curing agents in the U.S. About 6000 tons of anhydrides are used for this purpose. More than 220,000 tons of epoxy resins are used annually in the U.S.; worldwide capacity is 400,000 tons. The principal trade names are Araldite, Epon, Epirez, DER, and Epotuf.

| Table 7.6  Properties of Typical EP Plastics | | |
|---|---|---|
| **Property** | **Unfilled EP** | **Fiberglass-filled EP** |
| Heat-deflection temperature at 1.82 MPa, °C | 140 | 150 |
| Maximum resistance to continuous heat, °C | 120 | 135 |
| Coefficient of linear expansion, cm/cm/°C $\times 10^{-5}$ | 2.5 | 2.0 |
| Tensile strength, MPa | 52 | 83 |
| % Elongation | 5 | 4 |
| Flexural strength, MPa | 124 | 103 |
| Compressive strength, MPa | 70 | 100 |
| Notched Izod impact strength, J/m | 11 | 25 |
| Hardness, Rockwell | M90 | M105 |
| Specific gravity | 1.0 | 1.8 |

About 70,000 tons of EP are used annually as resin composites in the U.S., much of it by the electronics and transportation industries. While EP has a low order of toxicity and is not a carcinogen or mutagen, the amine curing agents may cause dermatitis if handled without protective gloves or protective clothing.

The properties of typical fiberglass-filled and unfilled EP resins are shown in Table 7.6.

## FURAN RESINS

Phenolic (PF) resin composites, which were the principal plastics prior to 1925, were attacked by alkalies. Hence, the development of alkaline-resistant furan resins was welcomed by those technologists who were using PF in alkaline environments. Since these resins were produced from furfuryl alcohol, which was obtained by the acid decomposition of corncobs, they were of particular interest to those who advocated chemurgy, i.e., agricultural sources of chemicals.

| Table 7.7 Properties of a Typical Carbon-Filled Furan Resin | |
| --- | --- |
| **Property** | |
| Heat-deflection temperature at 1.82 MPa, °C | 80 |
| Maximum resistance to continuous heat, °C | 100 |
| Coefficient of linear expansion, cm/cm/°C $\times 10^{-5}$ | 7.5 |
| Tensile strength, MPa | 41 |
| % Elongation | 1.5 |
| Flexural strength, MPa | 34 |
| Compressive strength, MPa | 69 |
| Notched Izod impact strength, J/m | 5 |
| Hardness, Rockwell | R110 |
| Specific gravity | 1.7 |

Furfuryl alcohol ($C_4H_3OCH_2OH$) is a cyclic alcohol which is resinified in acid media. The dark prepolymer has been used in polymer concrete, as a foundry core binder, and as an adhesive. Fiberglass-reinforced polymers of furfuryl alcohol have been used as corrosion-resistant structures. The properties of a typical carbon-filled polymer of furfuryl alcohol (Alkor) are shown in Table 7.7.

## PHENOLIC PLASTICS

Phenolic resins, which are produced at an annual rate of 1.4 million tons in the U.S., are the principal thermosets. Much of these resins is used as an adhesive for wood chips and panels and for fiberglass insulation. Annually, 740,000 tons of phenolic resins are used for plywood and 90,000 tons for molded products.

Commercial phenolic resins (Bakelite) were introduced by Leo Baekeland in the early 1900s, and his recipes for the reaction of phenol ($C_6H_5OH$) and formaldehyde (HCHO) continue to be used throughout the

Table 7.8  Properties of Typical Phenolic Plastics

| Property | Wood flour-filled | Mineral-filled | Fiberglass-filled |
|---|---|---|---|
| Heat-deflection temperature at 1.82 MPa, °C | 165 | 200 | 200 |
| Maximum resistance to continuous heat, °C | 160 | 175 | 175 |
| Coefficient of linear expansion, cm/cm/°C $\times 10^{-5}$ | 3.0 | 2.0 | 2.0 |
| Tensile strength, MPa | 48 | 90 | 41 |
| % Elongation | 0.5 | 0.5 | 0.5 |
| Flexural strength, MPa | 62 | 82 | 85 |
| Compressive strength, MPa | 85 | 90 | 200 |
| Notched Izod impact strength, J/m | 172 | 175 | 175 |
| Hardness, Rockwell | M100 | M110 | M110 |
| Specific gravity | 1.4 | 1.5 | 1.6 |

world. Under alkaline condition, this condensation-type polymerization produces resoles, and under acid conditions, novolacs.

The resole resins are prepolymers, which will crosslink (thermoset) on standing at room temperature or in the presence of a strong acid, such as *p*-toluene sulfonic acid ($H_3CC_6H_4SO_3H$). The resole resins are used for bonding plywood, fiberglass, and wood chips. The principal steps in this condensation are similar to those described for alkyd resins earlier in this chapter.

The novolac resins are linear polymers, which contain less than the stoichiometrical amounts of formaldehyde required for crosslinking. The additional formaldehyde is supplied by the addition of hexamethylene-tetramine (hexa($CH_2)_6N_4$), which is thermally decomposed to produce formaldehyde and ammonia in the molding cycle. PF molding powder may be reinforced with wood flour, carbon, or fiberglass. The properties of typical PF composites are shown in Table 7.8.

## UNSATURATED POLYESTERS

While alkyds are polyesters, the term "polyester" is commonly used to describe unsaturated polyester plastics and polyethylene terephthalate (PET) fibers. The unsaturated polyesters were developed by Carlton Ellis, who used an unsaturated anhydride (maleic anhydride, $(COCH)_2O$) instead of the saturated phthalic anhydride used in alkyds. Ellis also used a polymerizable solvent instead of xylene for the prepolymer and reinforced the brittle resins with fiberglass, whose commercialization coincided with Ellis' production of unsaturated polyesters.

As stated previously, the alkyds are produced by condensation (step reaction polymerization). Since Ellis used a difunctional alcohol (ethylene glycol), his prepolymers were linear, and the average degree of polymerization (DP) was predictable from Carothers' pioneer equation in which p is equal to the extent of polymerization.

Thus, if $p = 0.95$, the degree of polymerization or (n) would be 20. Hence, the chains in linear polymers would be too short to permit entanglement, and the prepolymer would not be useful as a plastic. However, Ellis dissolved this prepolymer in vinyl acetate ($H_2C= CHOOCCH_3$) and later replaced this monomer with styrene ($H_2C= CHC_6H_5$). The monomer was then polymerized in the presence of the unsaturated polyester prepolymer to produce a crosslinked polymer.

It is of interest to note that the prepolymer was produced by a step reaction polymerization and that the polymerization of the styrene was a free radical chain polymerization, as discussed previously for allyl polymers. In addition to the formation of an interpenetrating network by the second step, some grafting of the styrene chain occurs on the polyester backbone. A graft is a branch formed on a polymer chain, which in this case is polystyrene (PS).

It is customary to add a tertiary aliphatic amine such as dimethylaniline to accelerate the formation of free radicals from the benzoyl peroxide, as a result of an oxidation-reduction (redox) reaction. Driers, such as cobalt naphthenate, may also be added along with the BPO or methyl ethyl ketone peroxide (MEKP).

Other unsaturated prepolymers have also been used in place of the polyethylene maleate. One of these is vinyl ester, which is produced by the reaction of 2 moles of methacrylic acid with an epoxide of bisphenol A. Bisphenol A fumaric acid and isophthalic polyesters have also been

| Table 7.9  Properties of Typical Polyesters | | | | |
|---|---|---|---|---|
| Property | Unfilled | Chopped roving-filled | SMC | Fiber-glass-filled |
| Heat-deflection temperature at 1.82 MPa, °C | 79 | 180 | 180 | 200 |
| Maximum resistance to continuous heat, °C | 120 | 160 | 160 | 160 |
| Coefficient of linear expansion, cm/cm/°C $\times 10^{-5}$ | 5 | 2.5 | 2.5 | 2.5 |
| Tensile strength, MPa | 40 | 60 | 70 | 70 |
| % Elongation | 5 | 2 | 1 | 1 |
| Flexural strength, MPa | 80 | 100 | 100 | 85 |
| Compressive strength, MPa | 60 | 150 | 150 | 170 |
| Notched Izod impact strength, J/m | 20 | 200 | 250 | 200 |
| Hardness, Rockwell | M70 | M50 | M50 | M50 |
| Specific gravity | 1.2 | 1.6 | 2.2 | 2 |

used to produce prepolymers. The properties of typical polyester composites are shown in Table 7.9.

## POLYIMIDES

The first polyimide (PI) was synthesized by Bogart and Renshaw in 1908 when they thermally dehydrated 4-aminophthalic anhydride. The equation for this synthesis is:

However, since polymeric products such as PI's with aromatic units in the main chain are usually intractable, it is customary to first prepare a

Table 7.10  Properties of Typical Polyimides

| Property | Thermo-plastic PI | 50% glass-filled thermoset PI |
|---|---|---|
| Heat-deflection temperature at 1.82 MPa, °C | 315 | 350 |
| Glass transition temperature ($T_g$), °C | 300 | 310 |
| Maximum resistance to continuous heat, °C | 300 | 325 |
| Coefficient of linear expansion, cm/cm/°C $\times 10^{-5}$ | 5.0 | 1.3 |
| Tensile strength, MPa | 96 | 44 |
| % Elongation | 3 | 1 |
| Flexural strength, MPa | 172 | 145 |
| Compressive strength, MPa | 241 | 234 |
| Notched Izod impact strength, J/m | 100 | 300 |
| Hardness, Rockwell | E60 | M118 |
| Specific gravity | 1.4 | 1.6 |

polyamic acid and then dehydrocyclize (imidize) this soluble linear polyamic acid in a second step. Solutions of the polyamic acid precursor of aromatic PI's may be admixed with reinforcing fibers and imidized *in situ*. It is also possible to thermally dehydrocyclize polyimides capped by nonbornene or acetylene derivatives or to use nonstoichiometric quantities of bismaleimides and aromatic diamines to produce PI's.

However, since these techniques require the evaporation of high boiling solvents, such as dimethyl formamide (DMF) or *N*-methyl-2-pyrrolidone (NMP), *in situ*, high-temperature polymerization techniques are used to produce a thermooxidizable stable PI resin (PMR) which is used as a solution in aliphatic solvents.

PI film, which is solder resistant and dimensionally stable at elevated temperatures, has been used for printed circuits. Composites of PI with glass, carbon, or aramid fibers have been used for compressor valves in gas-cooled atomic reactors and in aircraft. PI composites have great po-

| Table 7.11  Properties of Typical BMI's |         |                   |                     |                     |
|-----------------------------------------|---------|-------------------|---------------------|---------------------|
| Property                                | BMI     | BMI/ 60% glass    | BMI/ 60% carbon     | BMI/ 70% carbon     |
| Tensile strength, MPa                   | 97      | ...               | 1725                | 570                 |
| Tensile modulus, GPa                    | 4.1     | ...               | 148                 | 16                  |
| Flexural strength, MPa                  | 210     | 480               | 2000                | 725                 |
| Coefficient of expansion, $10^{-5}/^\circ C$ | 3.1 | ...           | ...                 | ...                 |
| Maximum use temperature, °C             | 260     | 225               | 215                 | 175                 |

tential, and research and development in this field is continuing. The properties of typical polyimides are shown in Table 7.10.

Bismaleimides (BMI) are produced by the condensation of maleic anhydride and an aromatic diamine such as methylene dianiline. BMI composites are obtained by molding prepregs at about 210 °C for 2h, followed by a postcure of about 235 °C. The properties of typical BMI's are shown in Table 7.11.

## ELASTOMERS

While they are not recognized as thermosets by some technologists, soft and hard vulcanized rubber, which were produced by Charles and Nelson Goodyear, were the first crosslinked man-made polymers. As was evident by the limited use prior to the 1840s, natural rubber and other elastomers must be crosslinked (vulcanized, cured) to prevent stickiness and cold flow. These elastomers are described in Chapter 8.

## REFERENCES

- R.S. Bauer, Ed., *Epoxy Resin Chemistry*, ACS Symposium Series 114, American Chemical Society, Washington, DC, 1979
- P.F. Bruins, *Unsaturated Resin Technology*, Gordon and Breach Science Publishers, New York, 1976

- R. Burns, *Polyester Molding Compounds*, Marcel Dekker, New York, 1982
- M.A. Chaudhari, (Bismaleimides), *Modern Plastics*, Vol. 66 (No. 11), 1989, p. 142
- C.D. Dudgean, (Unsaturated polyesters), in *Engineered Materials Handbook*, Vol. 2, *Engineering Plastics*, ASM International, Metals Park, OH, 1988
- J.J. Fisher, (Amino plastics), in *Engineered Materials Handbook*, Vol. 2, *Engineering Plastics*, ASM International, Metals Park, OH, 1988
- M.R. Greer, (Allylic resins), in *Engineered Materials Handbook*, Vol. 2, *Engineering Plastics*, ASM International, Metals Park, OH, 1988
- H.J. Harrington, (Phenolic resins), in *Engineered Materials Handbook*, Vol. 2, *Engineering Plastics*, ASM International, Metals Park, OH, 1988
- J.A. Harvey, (Bismaleimide plastics), in *Engineered Materials Handbook*, Vol. 2, *Engineering Plastics*, ASM International, Metals Park, OH, 1988
- J.I. Kroschwitz, Ed., *Encyclopedia of Polymer Science and Engineering*, Vol. 4, Wiley-Interscience, New York, 1986
- G. Lubin, Ed., *Handbook of Composites*, Van Nostrand Reinhold, New York, 1982
- C.A. May and Y. Tamaka, *Epoxy Resin Chemistry and Technology*, Marcel Dekker, New York, 1973
- D.C. Miles and J.H. Briston, *Polymer Technology*, Chemical Publishing, New York, 1979
- T.C. Patton, *Alkyd Resin Technology*, Wiley-Interscience, New York, 1962
- J.S. Puglisi and M.A. Chaudhari, in *Engineered Materials Handbook*, Vol. 2, *Engineering Plastics*, ASM International, Metals Park, OH, 1988
- R.B. Seymour, *Plastics vs. Corrosives*, Wiley-Interscience, New York, 1982
- R.B. Seymour and G.S. Kirshenbaum, Ed., *High Performance Resins*, Elsevier Science, New York, 1986
- R.B. Seymour and R.F. Storey, Styrene-Alkyd Resins, *Eur. Coatings J.*, Vol. 430, 10 Sept 1990
- D.A. Shimp, (Cyanate esters), in *Engineered Materials Handbook*, Vol. 2, *Engineering Plastics*, ASM International, Metals Park, OH, 1988

- R.W. Tess and G.W. Poehlein, Ed., *Applied Polymer Science*, ACS Symposium Series 285, American Chemical Society, Washington, DC, 1985
- A. Whelan and J.A. Brydson, *Developments in Thermosetting Resins*, Applied Science Publishers, London, 1975
- A.A.K. Whitehouse, E.G.K. Pritchett, and G. Barnett, *Phenolic Resin Chemistry*, Plastics and Rubber Institute, London, 1967

# CHAPTER 8
# Rubbers

## ELASTICITY

All solid and liquid materials are elastic within reversible characteristic limits called the elastic limit. Extension beyond this limit is based on an irreversible dimensional change called plastic deformation. Reversible elasticity is uniquely associated with elastomers (rubbers), which are able to undergo a high order of stress-strain behavior without rupture. Thus, while the elastic limit of materials such as metals or glass is of the order of 3%, that of elastomers is of the order of 500%. Nevertheless, the Young's modulus (E) for rubber is only 0.06% that of polystyrene and 0.003% that of glass. However, Young's modulus of rubber is increased by the addition of carbon black in accordance with the Einstein-Guth-Gould theory in which $E_0$ is the modulus of the unfilled rubber and C is the concentration of the filler:

$$E = E_0 (1 + 2.5C + 14.1C^2)$$

The rubbery properties of natural rubber (*Hevea braziliensis*) and other elastomers at temperatures above $T_g$ are based on long-range elasticity. Unstretched elastomers are usually amorphous and consist of coils, with maximum entropy (S). When a force is applied, the chains tend to line up and irreversible flow will occur unless occasional crosslinks are present to restrict slippage of the chains. Vulcanized rubber consists of an average of one crosslink for every hundred carbon atoms. The chains between the crosslinks (principal sections) restrict cold flow but do not prevent uncoiling of the stretched chains, which return to the most probable conformation when the force is removed. If strain (elongation) is plotted against stress (load), the slope of the curve increases as the load increases but decreases when high loads are applied. The chains continue to uncoil

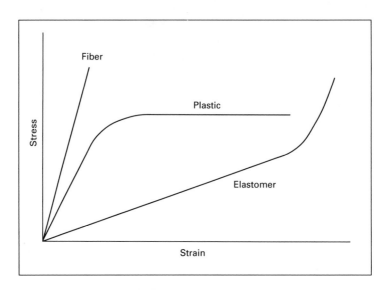

*Fig. 8.1 Typical stress-strain curves for different types of polymers*

prior to the attainment of maximum slope; i.e., there is a transition from random coils to an oriented state. The different stress-strain diagrams for fibers, plastics, and elastomers are shown in Fig. 8.1.

Elastomer molecules crystallize at the end point of elastic deformation, at which the molecules have a relatively high modulus (G) and low entropy (S). A much greater force is required for elongation after the chains have been aligned along the axis of elongation. The work done in the stretching process ($W_{el}$) is equal to the product of the retractile force (F) and the change in length (dl). Hence, the force is equal to the work per change in length (dl): $W_{el} = fdl$, or $f = W_{el}/dl$.

$W_{el}$ per change in length is equal to the change in energy (dG) per change in length, which is equal to the change in internal energy [(dE)/dl] minus the product of the change in entropy and the Kelvin temperature:

$$f = \frac{W_{el}}{dt} = \frac{dG}{dl} = \frac{dE}{dl} - \frac{Tds}{dl}$$

dE/dl is a measure of the energy absorbed in the original stretching process—i.e., before the end point of elastic deformation—and TdS/dl is a measure of the energy absorbed in the aligned molecular chains. Wohlisch and others have compared rubber-elastic deformation to the compression of a gas. For example, if one heats a stretched rubber band, there is a tendency for coil formation (i.e., increase in entropy) and then the tension increases much like the pressure of a gas, which increases when heated. In 1850, Kelvin and Clausius showed that the ratio of the retractile force (df) to the change in the temperature (dT) in an adiabatic process was equal to the specific heat of the elastomer ($C_p$) per degree Kelvin (T) times the change in temperature (dT) divided by the change in length (dl):

$$\frac{df}{dT} = \left(\frac{c_p}{T}\right)\left(\frac{dT}{dl}\right)$$

That an elastomer is cooled when stretched can be demonstrated qualitatively by placing a freshly stretched rubber band against the skin or lips. The presence of crystals in the stretched band may be observed by noting the opacity of a stretched band of gum rubber. Increase in temperature, when the stretched band is unstretched, can also be noted by pressing the unstretched band to the lips.

## ELASTOMER PRODUCTION DATA

While natural rubber (NR) accounts for only 29% of the total rubber consumption in North America and 34% worldwide, its annual consumption is equal to 90% of that of styrene-butadiene rubber (SBR) in the United States. In 1989, 2,824,000 tons of synthetic rubber (SR) and 980,000 tons of NR were consumed in the U.S. The worldwide consumption of SBR, polybutadiene (PB), ethylene propylene copolymer (EPDM), and acrylo-butadiene rubber (NBR) was 2.7 million tons, 580,000 tons, and 250,000 tons, respectively, in 1989. Annual production of butyl rubber and polychloroprene (neoprene) in the U.S. is 110,000 tons and 80,000 tons, respectively. Other synthetic elastomers (polyisoprene, silicone, etc.) accounted for 300,000 tons.

## ACRYLIC ELASTOMERS (ACM)

Copolymers of acrylic monomers, such as ethyl acrylate and butyl acrylate, are used as oil-resistant specialty elastomers. When reinforced with carbon black, these elastomers have a tensile strength of 15 MPa and an elongation of 215%. Other acrylic elastomeric copolymers are ethylene copolymers (PAMAC) and cyanoacrylic elastomers. Elastomer copolymers of acrylic ester and butadiene (ABR) are also available.

Vulcanized acrylic elastomers were produced by C.H. Fisher in 1944 and commercialized by Goodrich in 1948 under the trade name of Hylar PA. The softening points for polymethyl, ethyl, and *n*-butyl acrylate are −8, −20, and −40 °C, respectively. Polyacrylate elastomers are characterized by the following properties: tensile strength, 15 MPa; elongation, 280%; modulus at 100% elongation, 4.3 MPa; and Shore A hardness, 65.

## BUTYL RUBBER (IIR)

Since polyisobutylene is not readily crosslinked, Sparks and Thomas cationically copolymerized isobutylene with a small amount (5%) of isoprene. Most of the production of this copolymer (butyl rubber) is used by the tire industry. Unlike natural rubber and SBR, the green stretch (that of the unfilled elastomer) of butyl rubber is essentially equal to that of the carbon black-reinforced composite. Chlorobutyl rubber, which contains about 1.2% Cl, is more easily processed than butyl rubber, is more compatible with other elastomers, and is more resistant to gaseous permeation than other elastomers. IIR has a tensile strength of 18 MPa, an elongation of 700%, and a modulus at 400% elongation of 4 MPa.

## CHLOROSULFONATED POLYETHYLENES

Chlorosulfonated polyethylene (Hypalon) is produced by the reaction of a mixture of chlorine and sulfur dioxide on low-density polyethylene (LDPE). The processibility of this elastomer is a function of the chlorine content. The cured amorphous elastomer, which is resistant to ozone and sunlight, is used for wire coating and single-ply roofing.

## CIS-1,4-POLYBUTADIENE (BR)

*Cis*-isomers of both butadiene and isoprene are obtained when these dienes are polymerized by cobalt-containing catalysts or other coordination catalysts. The properties of both polybutadiene (Ameripol Tufsin) and polyisoprene (IR) are similar to those of natural rubber (*cis*-1,4-polyisoprene). The polyisoprene elastomers have been marketed under the trade names of Coral rubber and Cariflex.

IR has a glass transition temperature ($T_g$) of $-72$ °C, and the carbon black-filled IR has a tensile strength of 235 MPa and an elongation of 600%. The production of BR in the U.S. in 1989 was 400,000 tons.

## ETHYLENE-PROPYLENE COPOLYMERS (EPM/EPDM)

Copolymers of ethylene and propylene may be cured by heating with benzoyl peroxide. However, EPM polymers containing a small amount of a diene, such as dicyclopentadiene or norbornene (Nordel), are more readily cured by conventional techniques. These elastomers (EPDM) are unaffected by ozone and are used in the white sidewalls of pneumatic tires and in single-ply roofing. In 1989, 225,000 tons of EPDM were produced in the U.S. in 1989. EPDM has a tensile strength of 25 MPa, an elongation of 500%, a brittle point of 24 °C, a dielectric constant of 2.25, and a specific gravity of 0.86.

## POLYCHLOROPRENE (NEOPRENE, CR)

Polychloroprene is produced by the free radical chain emulsion polymerization of 2-chlorobutadiene (chloroprene). This elastomer (neoprene) is cured by heating with zinc oxide. The carbon black-filled polymer has properties similar to the "green strength" of the unfilled polymer. Neoprene is characterized by excellent resistance to oil at moderately elevated temperatures. Neoprene has a tensile strength of about 30 MPa and an elongation of 900% for the gum stock and 500% for the carbon-filled product.

## POLYSULFIDE ELASTOMERS (THIOKOL, T)

Polysulfide elastomers, which are produced by the condensation of alkylene dichlorides and sodium polysulfide, were synthesized by J. Baer in 1926 and J. Patrick in 1927. These solvent-resistant elastomers are cured by heating with zinc oxide. A typical carbon black (80 phr)-filled Thiokol has a tensile strength of 90 MPa and elongation of 260%. Much of the solid odoriferous Thiokol has been displaced by other oil-resistant elastomers. However, liquid Thiokols (LP-2, LP-3), which are produced by the reduction of the solid elastomer, are cured *in situ* and used as sealants and binders for rocket propellants.

## POLYETHER ELASTOMERS (ECH, CO)

Epichlorohydrin will copolymerize with ethylene oxide in the presence of an aluminum alkyl-water catalyst system in solution to produce an elastomer (Hydrin) which can be vulcanized to produce oil-resistant products. Vandenburg produced an amorphous copolymer from equal parts of epichlorohydrin and ethylene oxide. About 20,000 tons of ECH (Herchlor, Hydrin, Gechron) are produced annually. Nippon Zeon is the major producer of this oil-resistant polymer. Polyether block amide thermoplastic elastomers are also commercially available. Polymers of ethyl acrylate and chloroethyl vinyl ether (Lactoprene), which are crosslinked by heating with diamines, are used as oil-resistant gaskets.

## SILICONES (SIL)

Silicone elastomers, which consist of a backbone of siloxane (−Si−O−Si) repeating units with alkyl pendant groups, are produced by the hydrolysis of dialkylsilicon dihalides. These elastomers are cured by organic peroxy compounds and filled with amorphous silica. A typical filled and cured silicone elastomer will have a useful temperature range of −100 to 250 °C, a tensile strength of 70 MPa, and elongation of 300%. Silicone elastomers are used in many applications where resistance to high temperatures is essential.

## POLYPHOSPHAZENES

Polyphosphazenes are produced by heating hexachlorocyclotriphosphazene (phosphonitrilic chloride trimer) $(NPCl_2)_3$ at 250 °C. The polymer $(−[N=PCl_2−]_n)$ is readily hydrolyzed by water, but stable elastomers are obtained when the chlorine atoms are replaced by alkyl, amine, or trifluoroethoxy groups. Alkoxyphosphazene elastomers may be crosslinked by heating with the disodium salt of a diol.

Polyphosphazenes are stable over a temperature range of −50 to 200 °C. They are used as gaskets, pipe, and flexible sheet.

## FLUOROCARBON ELASTOMERS (FLU)

The fluorinated polyphosphazenes have a $T_g$ value of $-68$ °C, which is comparable to that of fluorosilicones. Organic peroxy curable fluoro-elastomers (Viton) are resistant to steam and many solvents. There are several fluorocarbon copolymer elastomers that are marketed under the trade name of Viton. These include copolymers of vinylidene fluoride/chlorotrifluoroethylene, tetrafluoroethylene/perfluormethyl vinyl ether (Kalrez), and vinylidene fluoride/hexafluoropropylene. The latter is also marketed under the trade name of Fluorel. In general, these fluorocarbon elastomers are useful over a wide temperature range ($-40$ to $315$ °C) and despite their high cost are used for solvent-resistant applications.

## POLYURETHANE ELASTOMERS (PUR)

Polyurethanes produced by the reaction of a diisocyanate with a poly-ester or polyether with terminal hydroxyl groups may be used as linear polymers or crosslinked by diamines or aminoalcohols. Some polyure-thane elastomers (Vulcolan) are processed and cured by techniques used for natural rubber. Others are cast from a two-component reactive system, i.e., reaction injection molding (RIM). Pneumatic tires and fillers for truck tires have been produced by the casting technique.

## POLYPENTENAMERS

Polypentenamers are produced by the coordination catalyzed, ring opening polymerization of cyclopentene at extremely high pressures (6500 MPa). *Trans*-polypentenamer, which has a $T_g$ of $-95$ °C and a melting point below room temperature, is obtained by the polymerization of cyclopentadiene. The polypentenamers may be cured by heating with sulfur at 170 °C.

## NATURAL RUBBER

*Cis*-polyisoprene has been isolated from over 200 different species of plants, e.g.: *Manihot glaziovii* in South America, *Castilloa elastica* in Mexico, *Landolphia* species and *Funtumia elastica* from Africa, *Tarax-acum koksaghyz* from Russian dandelions, *Cryptostegia grandiflora* from Floria milkweed, *Solidago* from goldenrod, *Parthenium argentatum*

(guayule) from the southwestern United States and northern Mexico, and
*Hevea braziliensis*, originally from South America.

*Hevea braziliensis*, which was smuggled from the Tapajoz region in
Brazil to Malaya via the Kew Gardens, near London, is the major source
of supply of natural rubber, but progress has been made in cultivating and
processing guayule rubber. More than 85% of all natural rubber is now
produced in Malaysia, Indonesia, Thailand, and Ceylon. Less than 1% of
NR is obtained from Brazil.

Rubber latex contains 30 to 40% NR and is obtained by slashing the
bark of rubber trees in order to reach the vertical latex vessels without
touching the cambium layer. Most of this latex is coagulated by the addition of formic acid (HCOOH) and dried as sheets, which are exposed to
smoke in an enclosed building at 40 to 60 °C. More than 80% of NR is
used as tires and mechanical goods.

Natural rubber and other elastomers can be compounded by adding
aqueous dispersions of additives to rubber latex, by drying and heat
curing, or by masticating the smoked sheet on a two-roll mill or in a
Banbury intensive mixer. In the so-called compounding of NR, the additives are added to the gummy mass. The last ingredient to be added by
the compounder is the accelerator.

The tensile strength of a typical vulcanized NR is 20 MPa at 680%
elongation. The Shore A hardness is about 42.

While *trans*-polyisoprene from *Sapotaceae* trees, gutta percha (*Palo-
quium oblongifolium*), and bully trees (balata, *Minusops globosa*) is a
high molecular weight polymer, it is not elastic. The elasticity of *cis*-
polyisoprene is dependent, to a large extent, on a crankshaft action of the
polymer backbone, which is not possible with the *trans*-polyisoprene.
Sketches of skeletal *cis* and *trans* forms are shown below:

*Cis* Isomer                    *Trans* Isomer

## STYRENE-BUTADIENE ELASTOMERS

Commercial elastomers called "methyl rubber" were produced by the
anionic polymerization of 2,3-dimethylbutadiene in the early 1900s. Tires

were made from "methyl rubber" during World War I, but they were not reinforced with carbon black and were not successful as replacements for NR.

A more useful elastomer was produced in the late 1920s by Tschunkur and Boch, who copolymerized butadiene and styrene (70:30) in an aqueous system. The elastomer was called Buna S, but there was no sodium (Na) used in the aqueous emulsion polymerization. Water-soluble compounds, such as potassium persulfate ($K_2S_2O_8$), are used as initiators for this copolymerization, which occurs within the soap micelles used as the emulsifying agent.

SBR, which was called GRS during World War II, was produced at relatively high temperatures, but a superior elastomer was obtained in the late 1940s by adding a redox system consisting of ferrous sulfate ($FeSO_4$ $\cdot 7H_2O$) with hydroperoxides and sodium formaldehyde sulfoxylate ($HOCH_2SO_2Na$).

The $T_g$ of SBR is $-52$ °C, but the tensile strength of SBR gum vulcanizate is only 2 MPa. However, when 50 phr of carbon black is added, the tensile strength increases to more than 20 MPa at 500% elongation. About 800,000 tons of SBR are used annually in the U.S. Most of this is used for tires and mechanical goods, which in many instances consist of blends of NR and SBR.

## ACRYLONITRILE-BUTADIENE ELASTOMERS

The elastomeric copolymers of acrylonitrile and butadiene (30:70) (Buna N, Perbunan, Chemigum, Hycar) were developed by Conrad and Tschunkur in the early 1930s. The solvent resistance and resistance to gaseous diffusion of these elastomers increases as the acrylonitrile ($CH_2=CHCN$) content is increased. These elastomers were used to produce the first synthetic rubber tire in the U.S., but now are used for gaskets and mechanical goods.

The composition of the copolymer (20 to 50% AN) has little effect on the tensile strength, but the copolymer with 40% AN has the highest tensile strength. These copolymers can be modified by the introduction of carboxylic acid ($CO_2H$) groups (COX). The tensile strength of vulcanized carbon black-filled COX (25 MPa) is more than 15% greater than NBR (21 MPa), and COX has a durometer hardness of 83 A vs. 65 A for NBR. The most outstanding difference between COX and NBR is abrasion resistance, measured by the Pico abrasion index (38 vs. 27). In 1989, 75,000 tons of NBR were produced in the U.S.

## THERMOPLASTIC ELASTOMERS (TPE)

Thermoplastic elastomers are readily processed and do not require crosslinking to achieve rubberlike properties. TPE's meet the ASTM definition for a rubber or elastomer, i.e.: "a material, which at room temperature, can be stretched repeatedly to at least twice its original length and upon release of stress returns immediately, with force, to its approximate original length." Since these products consist of both thermoplastic and elastomeric domains in the same macromolecule, reinforcement results from a thermally unstable network structure which exists at temperatures below the $T_g$ of the thermoplastic segments.

It should be emphasized that the blocks in TPE's are joined by strong covalent bonds and that TPE's are not mixtures or blends of two or more polymers. The Solprene TPE's, which have been called plastomers, consist of block copolymers with polybutadiene sequences, varying from 60 to 80%, joined to sequences or domains of polystyrene of 40 to 20%. Radial teleblock Solprene copolymers of isoprene and styrene (85:15) have a tensile strength of 15 MPa and elongation of 1200%.

## STYRENE-BUTADIENE BLOCK COPOLYMERS

Commercial TPE's based on block copolymers of styrene and butadiene or isoprene were introduced by Shell (Kraton) and Phillips Petroleum (Solprene) in the 1960s. Kraton is a block copolymer produced by the anionic sequential polymerization of styrene-butadiene-styrene (SBS) or styrene-isoprene-styrene (SIS). Hydrogenated SBS and SIB are also available. Because of the presence of the thermoplastic domains, the tensile strength of SIS is higher (21 MPa) than SBR (15 MPa), which is a random copolymer. A block copolymer of monomers A and B may be represented as $(A)_n(B)_n$ or $(A)_n(B)_n(A)_n$, while the random copolymer is (–ABAABABBA–).

SBS block copolymers, which are produced by Shell (Kraton), Firestone (Stereon), Asahi (Tufprene), and Enichen European (Solt), are used for wire and cable coatings, adhesives, and blends. These TPE's are readily extruded and injection molded.

## URETHANE-BASED TPE'S

Block copolymers of urethane are produced by condensation polymerization techniques similar to those described earlier for polyurethane elastomers. The TPE's based on blocks of polyesters and polyurethanes have higher strength than those based on blocks of polyethers, but they are less stable to hydrolysis. Polyester-polyurethane block copolymer fibers are marketed under the name of Spandex.

A typical polyester-polyurethane TPE (Estane, Pelethane, Cyanoprene) has a tensile strength of 35 MPa and an elongation of 500%. Both polyester and polyether polyurethane TPE's are marketed by Goodrich under the trade name of Vibrathane. A typical polyurethane TPE has a tensile strength of 30 MPa and an elongation of 500%. These products are characterized by exceptionally good abrasion resistance.

## POLYESTER-ETHER BLOCK COPOLYMERS

Segmented TPE's are produced by the condensation of polyesters and polyethers. A typical polyester-polyether TPE has a tensile strength of 38 MPa and an elongation of 450%. These products (Hytel) have excellent resistance to solvents and good tear resistance.

## POLYPROPYLENE/ETHYLENE-PROPYLENE COPOLYMER BLENDS

As stated in Chapter 6, TPE's can be produced by blending polypropylene with ethylene-propylene copolymers. These products, which are marketed under the trade names of Telcar, Somel, and Santoprene, have tensile strengths of 11.5 MPa for Somel 301 and 103, and 50 MPa for Santoprene. These TPE's have elongations of 230 and 375%, respectively.

## THERMOPLASTICS

While the pioneer thermosets and elastomers were all used as reinforced composites, most of the pioneer thermoplastics have been used without fillers. However, filled thermoplastics are now commercially available and will be discussed in Chapter 9.

# REFERENCES

- A. Barker, (Production data), *Elastomerics*, Vol. 122 (No. 2), 1990, p. 18
- A.K. Bhowmick and H.L. Stephens, *Handbook of Elastomers*, Marcel Dekker, New York, 1988
- C.C. Davis and J.T. Blake, *The Chemistry and Technology of Rubber*, Reinhold, New York, 1937
- H. Ehrend and T.L. Leo, (Specialty elastomers), *Elastomerics*, Vol. 122 (No. 5), 1990, p. 21
- W. Hofmann, *Rubber Technology Handbook*, Hansen Publishers, Munich, 1989
- J.I. Kroschwitz, Ed., *Encyclopedia of Polymer Science and Engineering*, Wiley-Interscience, New York, 1986
- M. Morton, Ed., *Rubber Technology*, Van Nostrand Reinhold, New York, 1973
- J.H. Saunders and K.C. Frisch, *Polyurethanes: Chemistry and Technology*, Interscience, New York, 1972
- R.B. Seymour, Ed., *History of Polymer Science and Technology*, Marcel Dekker, New York, 1982
- R.B. Seymour and G.S. Kirshenbaum, Ed., *High Performance Polymers: Their Origin and Development*, Elsevier Science Press, New York, 1986
- G.G. Winspear, *Rubber Handbook*, R.T. Vanderbilt Co., New York, 1968
- S. Wolkenbreit and B.M. Walker, Ed., *Handbook of Thermoplastic Elastomers*, Van Nostrand-Reinhold, New York, 1979

# CHAPTER 9

# General-Purpose Thermoplastics

## INTRODUCTION

As stated in Chapter 7, with reinforced plastic composites, the early emphasis was on reinforced thermosets. The lack of emphasis on reinforced thermoplastics was due, in part, to the inherent toughness of some of the pioneer thermoplastics, such as cellulose nitrate and cellulose acetate, the low cost of the general-purpose thermoplastics, and the mistaken concept that crosslinking was essential for reinforcement.

It is now recognized that crystalline thermoplastics are upgraded more than amorphous thermoplastics by the addition of fillers. Nevertheless, the first commercial reinforced thermoplastic (RTP) was fiberglass-reinforced polystyrene, which was introduced in the early 1950s by Fiberfill Inc. This was followed by the marketing of reinforced nylon 66 in the 1960s. The properties of this crystalline plastic are readily upgraded by the addition of fiberglass, and nylon 66 has been the leading RTP.

## ACRYLICS

The development of acrylic polymers by O. Haas in the early 1900s paralleled that of Bakelite. However, Haas was not able to commercialize these transparent amorphous plastics until the mid-1920s.

Alkyl esters of acrylic and methacrylic acids are readily polymerized by free radical chain reaction polymerization. The early emphasis was on the flexible polyalkyl acrylates. However, a harder copolymer of methyl acrylate and ethyl methacrylate was used as an inner layer in safety glass in the early 1930s; this led to the commercial production of the methacrylate polymers.

| Table 9.1 Properties of a Typical PMMA Sheet | |
|---|---|
| **Property** | |
| Heat-deflection temperature at 1.82 MPa, °C | 96 |
| Maximum resistance to continuous heat, °C | 90 |
| Coefficient of linear expansion, cm/cm/°C $\times 10^{-5}$ | 7.6 |
| Tensile strength, MPa | 72 |
| % Elongation | 5 |
| Flexural strength, MPa | 110 |
| Compressive strength, MPa | 124 |
| Notched Izod impact strength, J/m | 74 |
| Hardness, Rockwell | M93 |
| Specific gravity | 1.19 |

The use of acrylic copolymers in safety glass led to an investigation of the inner layer as a self-supporting glazing material. The original sheets (Plexiglas, Lucite) were produced by casting the catalyzed prepolymer; today, these sheets are also being produced by extrusion. Both unfilled and filled polymethyl methacrylate (PMMA) can be injection molded, but since the emphasis is on the inherent clarity of the polymer, there is not much interest in filled acrylic polymers.

The principal application of PMMA is as sheet for glazing and signs. Acrylic polymer emulsions and solutions are used as coatings for metal and wood surfaces, textiles, and paper. Polymers of higher molecular weight alkyl acrylates are used as oil additives. PMMA is also used for biomedical applications, optical fibers, and cultured marble. More than 335,000 tons of acrylic plastics are used annually in the U.S.; worldwide production capacity is in excess of 1.3 million tons. The properties of a typical PMMA sheet are shown in Table 9.1.

## FLUOROCARBON PLASTICS

Fluorocarbon polymers are high-performance plastics, but they lack the load-bearing properties of engineering plastics. They are characterized by good impact resistance over a wide temperature range, excellent electrical properties, low coefficient of friction, and outstanding resistance to corrosives.

In much of polymer technology, science has lagged behind usage because of a disregard for this science by many organic chemists and acceptance of erroneous concepts by some polymer scientists. Most polymer scientists believed that vinyl fluoride ($H_2C=CHF$) would not form a polymer and that substitution of all ethylenic hydrogen atoms in the monomer tetrafluoroethylene ($F_2C=CF_2$, TFE) would prevent polymerization of that monomer. Hence, the development of fluorocarbon plastics was delayed until the late 1930s.

As many scientists and nonscientists know, TFE stored in a cylinder *does* polymerize spontaneously, and the chemist who investigated TFE as a refrigerant found that the polymer (Teflon, PTFE) was the most important end product for this gas. Roy Plunkett patented PTFE, but because of the ability of PTFE to handle the corrosives used in the preparation of uranium isotopes for atomic energy applications, widescale use was delayed until after World War II.

PTFE is now produced by the free radical chain polymerization of TFE. Because of the large volume occupied by the closely packed pendant fluorine atoms, PTFE exists as a stiff polymeric chain that cannot be plasticized and cannot be processed by conventional techniques. PTFE is processed by sintering techniques used by the metals industry; i.e., a preform is heated at 327 °C to coalesce the infusible particles.

The more readily processible polychlorotrifluoroethylene (PCTFE) was polymerized by free radical chain polymerization in Germany in 1934 and in the U.S. in the late 1930s. While its lubricity is slightly lower than that of Teflon, films of PCTFE (Kel F) have outstanding vapor resistance.

Polyvinylidene fluoride (PVDF, Kynar), which has a lower specific gravity and a higher coefficient of friction than PCTFE, is melt processible. In addition to being used as a corrosion-resistant film and coating, PVDF is used in piezoelectric applications.

Polyvinyl fluoride (PVF, Tedlar) is a chemically inert plastic that is readily fabricated into tough weather-resistant films, fibers, coatings, and

| Table 9.2  Properties of Typical Polyfluorocarbon Plastics | | | | |
|---|---|---|---|---|
| **Property** | **PTFE** | **PCTFE** | **PVDF** | **PVF** |
| Heat-deflection temperature at 1.82 MPa, °C | 100 | 100 | 80 | 90 |
| Maximum resistance to continuous heat, °C | 250 | 200 | 150 | 125 |
| Coefficient of linear expansion, cm/cm/°C $\times 10^{-5}$ | 10 | 14 | 8.5 | 10 |
| Tensile strength, MPa | 24 | 34 | 55 | 65 |
| Flexural strength, MPa | 50 | 60 | 75 | 90 |
| Notched Izod impact strength, J/m | 160 | 100 | 150 | 100 |
| % Elongation | 200 | 100 | 200 | 200 |
| Hardness | D52(a) | R80(b) | R110(b) | R83(b) |
| Specific gravity | 2.16 | 2.1 | 1.76 | 1.4 |
| Dielectric constant | 2.1 | 3.0 | 6.1 | ... |
| (a) Shore hardness. (b) Rockwell hardness | | | | |

moldings. In addition to these polymers with varying numbers of pendant fluorine atoms, there are many copolymers of these monomers and other vinyl compounds. The properties of typical fluorocarbon plastics are shown in Table 9.2. The effect of fillers on the properties of PTFE is shown in Table 9.3.

## POLYOLEFINS

Paraffin, which is an olefin oligomer, has been known for centuries. However, its molecular weight is below the threshold required for chain entanglement and the intermolecular forces are too low to counteract this deficiency. However, information on the chemical resistance of higher molecular weight polyolefins and on crystallinity can be obtained from a study of a straight chain oligomer, such as $n$-decane $(H(CH_2)_{10}H)$.

Several scientists had synthesized linear polyethylene by the decomposition of diazomethane in the early 1900s, and Carl Marvel actually pro-

Table 9.3  Properties of Typical Filled PTFE

| Property | Unfilled PTFE | 15% glass | 25% glass | 15% graphite | 60% bronze |
|---|---|---|---|---|---|
| Thermal conductivity, mW/MK | 0.244 | 0.37 | 0.45 | 0.45 | 0.45 |
| Tensile strength, MPa | 28 | 25 | 17.5 | 21 | 124 |
| % Elongation | 350 | 300 | 250 | 250 | 150 |
| Notched Izod impact strength, J/m | 152 | 146 | 119 | 100 | 75 |
| Coefficient of friction, 3.4 MPa load | 0.08 | 0.13 | 0.13 | 0.10 | 0.10 |
| Wear factor, 1/pPa | 5013 | 280 | 26 | 102 | 12 |
| Shore durometer hardness | 51D | 54D | 57D | 61D | 70D |
| Specific gravity | 2.18 | 2.21 | 2.24 | 2.16 | 3.74 |

duced high-density polyethylene (HDPE) in the early 1930s by the polymerization of ethylene in the presence of a coordination catalyst. Despite the existence of these polymers, no commercialization occurred until 1939, when ICI marketed low-density polyethylene (LDPE), which had been produced serendipitously under extremely high pressure (200 MPa) by Fawcett and Gibson in 1933.

Fortunately, there is a large family of polymers and copolymers of olefins; these will be discussed below. It is important to note that the effect of branching on density and of regular structure on crystallizability as well as the lack of resistance to weathering of polymers with tertiary hydrogen atoms is readily demonstrated by specific polyolefins.

## Low-Density Polyethylene

Since this partially crystalline (50%) polymer is produced by the free radical chain polymerization of ethylene, under extremely high pressures, it was originally called high-pressure polyethylene (HPPE). The original initiator was a trace of oxygen, but it is now standard practice to use organic peroxides as initiators. Nevertheless, Nobel laureate Hermann Staudinger (1953) refused to believe that ICI chemists had polymerized

Table 9.4  Properties of a Typical LDPE Plastic

| Property | |
| --- | --- |
| Glass transition temperature $(T_g)$, °C | −25 |
| Coefficient of linear expansion, cm/cm/°C $\times 10^{-5}$ | 15 |
| Tensile strength, MPa | 20 |
| % Elongation | 350 |
| Shore hardness | 47D |
| Specific gravity | 0.925 |

ethylene ($H_2C=CH_2$), for he maintained that such a polymerization was impossible.

Despite the hazards associated with this high-pressure polymerization and the availability of alternatives, more than 3.3 million tons of LDPE were produced in the U.S. in 1989. The properties of a highly branched, translucent LDPE are shown in Table 9.4.

## High-Density Polyethylene

In 1951, Larcher and Pease used extremely high pressure (2000 MPa) for the free radical chain polymerization of ethylene and obtained polymers with high specific gravities (0.96), which indicated that their product was less highly branched than LDPE and could be considered to consist of linear chains, which occupy less volume than branch chains. This product had a higher melting point (127 °C) and was more crystalline than LDPE (55%).

Several other contemporary researchers, such as Zletz, Hogan and Banks, and Ziegler also polymerized ethylene at moderate pressures using a cobalt molybdate catalyst ($CoMoO_4$) on alumina ($Al_2O_3$), chromic oxide ($CrO_3$) on silica ($SiO_2$), or aluminum trialkyl ($AIR_3$) and titanium tetrachloride ($TiCl_4$), respectively, and obtained linear polyethylene. All these researchers were awarded U.S. patents; Ziegler was awarded the Nobel Prize in 1963.

Table 9.5  Properties of Typical Filled and Unfilled HDPE

| Property | HDPE | 30% glass-filled HDPE |
|---|---|---|
| Melting point ($T_m$), °C | 130 | 140 |
| Heat-deflection temperature at 1.82 MPa, °C | 40 | 120 |
| Maximum resistance to continuous heat, °C | 40 | 110 |
| Coefficient of linear expansion, cm/cm/°C $\times$ $10^{-5}$ | 10 | 5 |
| Tensile strength, MPa | 27 | 62 |
| % Elongation | 110 | 1.5 |
| Flexural strength, MPa | ... | 76 |
| Compressive strength, MPa | 21 | 43 |
| Notched Izod impact strength, J/m | 133 | 64 |
| Hardness | D40(a) | R75(b) |
| Specific gravity | 0.95 | 1.3 |

(a) Shore hardness. (b) Rockwell hardness

Since this high-density polymer (0.96 g · cm$^{-3}$) was made at relatively low pressures, it was originally called low-pressure polyethylene (LPPE). The Phillips (Hogan and Banks) process is the major one now used for the production of HDPE. More than 3.6 million tons of HDPE were produced in the U.S. in 1989, and more than 1.3 million tons are used annually in the U.S. for the blow molding of containers such as bottles, drums, and tanks. About 700,000 tons of HDPE are injection molded as crates, housewares, and packages; more than 815,000 tons are extruded as pipe, rope, and netting. Some of the filaments are chain extended to form extremely strong filaments (Spectra).

The extent of branching, as measured by the number of methyl (CH$_3$) groups per 100 carbon atoms or per 50 mers, is lowest for the high-density polymer (0.96 g · cm$^{-3}$) and decreases as density and crystallinity increase. The crystallinity of the highest density HDPE may be as much as 95%, while that of LDPE is as low as 50%.

As shown in Table 9.5, the addition of 30% fiberglass results in a dramatic improvement in the physical properties of HDPE.

## Ultrahigh Molecular Weight HDPE (UHMWPE)

Since most useful properties of polymers level off after a threshold molecular weight has been reached, it is customary to keep the molecular weight within the so-called commercial range to avoid the additional costs of processing melts with increasing viscosities. Nevertheless, some properties, such as toughness, continue to increase as the molecular weight increases, and the high costs are justified when HDPE is produced for trash barrels, etc. The impact strength of UHMWPE is much higher than that of HDPE. The tensile strength is also higher and could be increased further by the addition of fiberglass. Nevertheless, it is not practical to reinforce UHMPE, since reinforcement reduces the toughness as measured by impact tests.

## Linear Low-Density Polyethylene (LLDPE)

While it is difficult to produce copolymers of ethylene and alpha olefins using the ICI high-pressure process, copolymers can be produced readily by the Phillips or Ziegler processes. In the early 1970s in Canada, DuPont produced low-density linear copolymers of ethylene by adding higher molecular weight alpha olefins, such as 1-hexene, to the reactants. It is of interest to note that while the Phillips process is the one used to produce most of the world's HDPE, LLDPE is made by a modified Ziegler process.

The principal LLDPE process is the Union Carbide Unipol low-pressure (20 MPa) gas-phase fluidized-bed process, in which copolymers of ethylene with about 8% 1-butene, 1-hexene, or 1-octene are produced in a continuous process. Annual U.S. production of LLDPE is now 1.5 million tons, and the volume of this tough copolymer is growing rapidly. The many different types of polyethylene have been classified in accordance with their densities (g · cm$^{-3}$) (Table 9.6). The properties of a typical LLDPE plastic are shown in Table 9.7. A very low density polyethylene (VLDPE) (0.890 to 0.915 g · cm$^{-3}$) is also commercially available.

| Table 9.6  Classification of Polyethylenes | |
|---|---|
| **Type** | **Density, g · cm$^{-3}$** |
| I | 0.910-0.095 |
| II | 0.926-0.940 |
| III | 0.941-0.959 |
| IV | ≤0.960 |

| Table 9.7  Properties of a Typical LLDPE Plastic | |
|---|---|
| **Property** | |
| Melting point (T$_m$), °C | 122 |
| Tensile strength, MPa | 20 |
| % Elongation | 400 |
| Shore hardness | D55 |
| Specific gravity | 0.93 |

## Ethylene Copolymers

In addition to the copolymer of ethylene and propylene (EPM and EPDM) discussed in Chapter 8, commercial copolymers of ethylene and vinyl acetate (EVA), vinyl alcohol (EVAL), ethyl acrylate, and methacrylic acid (ionomers, Surlyn) are commercially available.

Random copolymers of ethylene with 5 to 50% vinyl acetate are produced at high pressure using ICI techniques. These amorphous copolymers have a high degree of clarity and are used in packaging. These copolymers may be hydrolyzed to produce vinyl alcohol copolymers, which have excellent resistance to gaseous diffusion. The properties of typical copolymers of ethylene/vinyl acetate and ethylene/vinyl alcohol are shown in Table 9.8.

Table 9.8  Properties of Typical EVA and EVAL Copolymers

| Property | EVA | EVAL |
|---|---|---|
| Heat-deflection temperature at 1.82 MPa, °C | 65 | ... |
| Coefficient of linear expansion, cm/cm/°C $\times 10^{-5}$ | 16 | ... |
| Tensile strength, MPa | 27 | 65 |
| % Elongation | 500 | 230 |
| Shore hardness | D30 | ... |
| Specific gravity | 0.93 | 1.16 |

Table 9.9  Properties of a Typical Ionomer

| Property | |
|---|---|
| Heat-deflection temperature at 1.82 MPa, °C | 40 |
| Coefficient of linear expansion, cm/cm/°C $\times 10^{-5}$ | 15 |
| Tensile strength, MPa | 27 |
| % Elongation | 400 |
| Shore hardness | D60 |
| Specific gravity | 0.95 |

Copolymers of ethylene and methacrylic acid (ionomers) were introduced commercially by DuPont in the 1950s under the trade name of Surlyn. Ionic crosslinks, which are thermally unstable, may be introduced by neutralizing ionomers with sodium or zinc. These ionic forces suppress the formation of large crystallites and hence reduce haze and enhance melt viscosity, which facilitates processing of film. Ionomer films are characterized by excellent resistance to permeation by vegetable oils over a wide temperature range and excellent resistance to abrasion. These resins may be filled with fiberglass and modified by the addition of elastomers. The properties of a typical high-stiffness ionomer are shown in Table 9.9.

## Polypropylene (PP)

Despite his genius as a researcher, Nobel laureate Karl Ziegler failed to realize that polypropylene had been produced but not characterized by his associate H. Breil at Mulheim. In addition, Bernard Wright of Petrochemicals Ltd. had produced large quantities of polypropylene using Ziegler's catalyst, but mistakenly believed that this polymer was included in the Ziegler patent.

Probably more important, Hogan and Banks, at Phillips Petroleum Company, had produced crystalline polypropylene with the chromic oxide/silica catalyst before Ziegler had produced HDPE. Nevertheless, Paolo Chini, an associate of Giulio Natta, produced stereospecific polypropylene on March 11, 1954, and Dr. Natta shared the Nobel prize with K. Ziegler in 1963 for this discovery. Yet, after some 20 years of litigation, the U.S. Patent Office declared that Hogan and Banks were the true inventors of crystalline polypropylene and issued patent No. 4,376,851 to the Phillips researchers on March 15, 1983.

Propylene was polymerized by cationic chain polymerization by Butlerov in the 1870s; this amorphous polymer was more thoroughly investigated by Fontana of Socony-Mobil Company in the 1950s. However, this amorphous product differed from the crystalline polymer obtained by Hogan and Banks and Natta. Mrs. Natta coined the names isotactic, syndiotactic, and atactic for the stereo isomers shown below; the commercial crystalline polypropylene is the isotactic isomer.

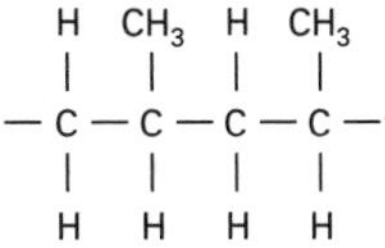

Isotactic PP

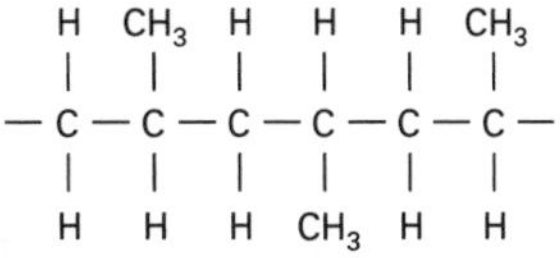

Syndiotactic PP

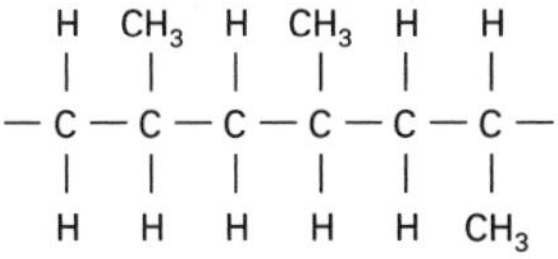

Atactic PP

Table 9.10  Properties of Typical Filled and Unfilled Polypropylene

| Property | Unfilled PP | 40% talc/ PP | 40% CaCO₃/ PP | 40% glass/ PP | 30% graphite/ PP |
|---|---|---|---|---|---|
| Melting point ($T_m$), °C | 170 | 168 | 168 | 163 | 168 |
| Heat-deflection temperature at 1.82 MPa, °C | 55 | 100 | 80 | 160 | 120 |
| Maximum resistance to continuous heat, °C | 100 | 120 | 110 | 135 | 125 |
| Coefficient of linear expansion, cm/cm/°C $\times 10^{-5}$ | 9 | 6 | 4 | 3 | 3 |
| Tensile strength, MPa | 35 | 32 | 26 | 82 | 47 |
| % Elongation | 150 | 5 | 15 | 2 | 0.5 |
| Flexural strength, MPa | 48 | 60 | 45 | 100 | 62 |
| Compressive strength, MPa | 45 | 52 | 35 | 64 | 55 |
| Notched Izod impact strength, J/m | 42 | 27 | 42 | 90 | 56 |
| Hardness, Rockwell | R90 | R100 | R88 | R105 | R100 |
| Specific gravity | 0.90 | 1.25 | 1.23 | 1.22 | 1.04 |

Most of the 3.3 million tons of polypropylene produced in the U.S. in 1989 were made using a modification of the Ziegler catalyst enhanced by the addition of magnesium chloride ($MgCl_2$). Propylene can also be produced using the Shell catalyst in the Unipol gas-phase fluidized-bed process. The increased efficiency of modern catalysts has eliminated the need to remove spent catalyst from this polymer.

Because of its high degree of crystallinity, the properties of polypropylene are enhanced dramatically by the addition of fillers and reinforcements. The properties of typical filled and unfilled polypropylene are shown in Table 9.10.

In addition to the elastomeric EPDM described in Chapter 8, there are several other commercial copolymers, such as the block copolymer of propylene and ethylene and numerous blends of polypropylene and other polymers.

### Polybutenes (PB)

Isobutylene (($CH_3)_2C=CH_2$) has been polymerized for several decades by cationic polymerization to produce a sticky solid. However, since the undesirable cold flow of polyisobutylene (PIB) was eliminated by copolymerization with isoprene to produce butyl rubber, production of PIB has been reduced in favor of the production of butyl rubber.

1-butene ($CH_2=CH-CH_2-CH_3$) is readily polymerized by Ziegler or Phillips catalysts to produce a tough isotactic polymer (Duraflex). The large-scale use of this polymer is hampered by a gradual irreversible phase transition from a metastable form II to a stable form I over a period of about one week. PB is extruded as pipe and is also used as a hot-melt adhesive. PB is compatible with other polyolefins, such as polypropylene. Since it may be loaded with as much as 75% filler, it is also used as a concentrate carrier for other additives. The principal use of 1-butene is a comonomer of LLDPE.

### Polymethylpentene (TPX)

4-methyl-1-pentene ($H_2C=CHCH_2C(CH_3)_2$ was polymerized by use of a $Cr^{6+}$ catalyst by Hogan and Banks in the 1950s. This transparent, high-melting isotactic polymer, which has the lowest specific gravity of any solid commercial polymer (0.83), is produced by Mitusi Petrochemical Industries in Tokyo. Its specific gravity is close to the lowest theoretical limit for noncellular thermoplastics.

Because of the presence of a tertiary hydrogen atom, TPX, like polypropylene, must be stabilized before being used outdoors or at elevated temperatures. TPX is injection molded and blow molded to produce shatter-resistant laboratory and medical ware. The properties of a typical polymethylpentene are shown in Table 9.11.

## POLYSTYRENE (PS)

Despite the fact that polystyrene ($-CH_2CH(C_6H_5)-$) had been produced by Simon in 1839, almost a century passed before this important plastic was produced commercially in Germany and the United States. Styrene is one of the few monomers that can be polymerized by all types of chain-reaction polymerization. However, most of this polymer is produced

Table 9.11  Properties of a Typical
Polymethylpentene

| Property | |
| --- | --- |
| Melting point ($T_m$), °C | 230 |
| Heat-deflection temperature at 1.82 MPa, °C | 50 |
| Maximum resistance to continuous heat, °C | 40 |
| Coefficient of linear expansion, cm/cm/°C $\times 10^{-5}$ | 6.5 |
| Tensile strength, MPa | 18 |
| % Elongation | 75 |
| Notched Izod impact strength, J/m | 133 |
| Hardness, Rockwell | R50 |
| Specific gravity | 0.83 |

Table 9.12  Properties of a Typical Styrene Polymer

| Property | PS | PS/ 30% glass |
| --- | --- | --- |
| Heat-deflection temperature at 1.82 MPa, °C | 90 | 105 |
| Maximum resistance to continuous heat, °C | 75 | 95 |
| Coefficient of linear expansion, cm/cm/°C $\times 10^{-5}$ | 7.5 | 4.0 |
| Tensile strength, MPa | 41 | 82 |
| % Elongation | 1.5 | 1.0 |
| Flexural strength, MPa | 83 | 117 |
| Compressive strength, MPa | 90 | 103 |
| Notched Izod impact strength, J/m | 21 | 20 |
| Hardness, Rockwell | M65 | M70 |
| Specific gravity | 1.04 | 1.2 |

by bulk or mass polymerization via organic peroxide-initiated chain-reaction polymerization. In 1989, 2.35 million tons of PS were produced in the U.S. Much of this production was high impact PS (HIPS) and foamed PS (EPS).

Polystyrene is an amorphous, clear, brittle polymer that can be injection molded, extruded, and blown to produce containers. The principal applications of polystyrene are packaging, serviceware, electronics, furniture, and appliances. The properties of polystyrene are shown in Table 9.12.

In addition to HIPS, discussed in Chapter 6, copolymers of styrene and acrylonitrile (SAN), maleic anhydride (SMA), methyl methacrylate (SMMA), and acrylonitrile/butadiene (ABS) are commercially available. The properties of these unfilled and filled copolymers are shown in Table 9.13.

## VINYL CHLORIDE POLYMERS AND COPOLYMERS

A polymer of vinyl chloride (VCM), like that of styrene, was produced in the 1830s. However, the product that was considered to be polyvinyl chloride (PVC) by Regnault may have in fact been polyvinylidene chloride (PVDC) ($-CH_2CCl_2-$). Regardless, commercial production was delayed until Waldo Semon and co-workers produced a plasticized PVC (Koroseal) in the early 1930s.

Subsequently, German chemists developed techniques for processing heat-stabilized PVC, and this rigid PVC was then available for extrusion and injection molding. PVC can be produced by peroxy-initiated chain polymerization using bulk, suspension, solution, or emulsion techniques. Commercial PVC is an amorphous atactic isomer with less than a 10% degree of crystallinity. When heated, PVC tends to degrade by loss of hydrogen chloride, but this zipperlike degradation is retarded in the presence of appropriate heat and UV stabilizers.

The $T_g$ of PVC (75 °C) may be increased by post-chlorination (CPVC). The increase of $T_g$ values is related to the percent chlorine added; e.g., a polymer with 57% chlorine will have a $T_g$ value of 95 °C and that with 70% chlorine will have a $T_g$ value of 150 °C. Figure 9.1 shows a pump made from CPVC. Most PVC is used without large amounts of filler. However, PVC floor tile is usually reinforced with asbestos.

Table 9.13  Properties of Typical Styrene Copolymers

| Property | SAN | SAN/ 20% glass | SMA | SMA/ 20% glass | SMMA | ABS | ABS/ 20% glass |
|---|---|---|---|---|---|---|---|
| Heat-deflection temperature at 1.82 MPa, °C | 100 | 105 | 107 | 112 | 100 | 95 | 105 |
| Maximum resistance to continuous heat, °C | 85 | 100 | 100 | 110 | 90 | 90 | 100 |
| Coefficient of linear expansion, cm/cm/°C × $10^{-5}$ | 7 | 3 | 8 | 4 | 5 | 8 | 2 |
| Tensile strength, MPa | 75 | 115 | 45 | 65 | 60 | 35 | 80 |
| % Elongation | 3 | 1.5 | 6 | 3 | 2.5 | 40 | 3 |
| Flexural strength, MPa | 90 | 145 | 85 | 120 | 105 | 70 | 100 |
| Compressive strength, MPa | 100 | 135 | ... | ... | ... | 50 | 95 |
| Notched Izod impact strength, J/m | 26 | 55 | 50 | 90 | 20 | 150 | 60 |
| Hardness, Rockwell | M80 | M95 | R107 | R73 | M75 | R80 | R107 |
| Specific gravity | 1.07 | 1.3 | 1.06 | 1.2 | 1.1 | 1.05 | 1.2 |

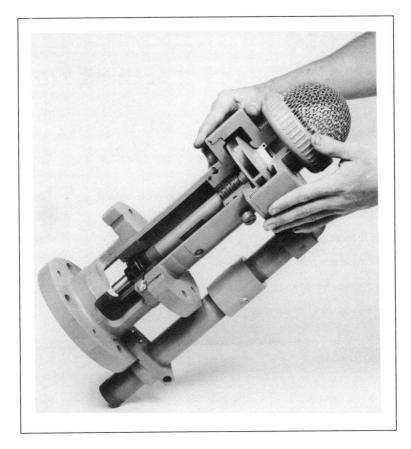

*Fig. 9.1 Pump made from CVPC (B.F. Goodrich)*

The most widely used copolymer of vinyl chloride is the one with vinyl acetate ($H_2C=CHOOCCH_3$) (Vinylite). The properties are comparable to those of PVC with moderate amounts of plasticizers. These copolymers are used in protective coatings. The properties of typical vinyl polymers are shown in Table 9.14. In 1989, 3.8 million tons of PVC were produced in the U.S.

## HIGH-PERFORMANCE POLYMERS

As stated in this chapter, the use of reinforced general-purpose thermoplastics is limited to a few applications, such as under-the-hood applications. However, reinforced high-performance polymers, which are

Table 9.14   Properties of Typical Vinyl Polymers

| Property | PVC | PVC/ 20% glass | CPVC |
|---|---|---|---|
| Heat-deflection temperature at 1.82 MPa, °C | 65 | 80 | 105 |
| Maximum resistance to continuous heat, °C | 55 | 75 | 100 |
| Coefficient of linear expansion, cm/cm/°C $\times 10^{-5}$ | 7 | 3 | 7 |
| Tensile strength, MPa | 60 | 75 | 60 |
| % Elongation | 30 | 4 | 100 |
| Flexural strength, MPa | 90 | 125 | 100 |
| Compressive strength, MPa | 95 | ... | 100 |
| Notched Izod impact strength, J/m | 50 | 80 | 50 |
| Hardness | D75(a) | D85(a) | R120(b) |
| Specific gravity | 1.4 | 1.5 | 1.5 |

(a) Shore hardness. (b) Rockwell hardness

described in Chapter 10, are widely used in many more demanding applications.

## REFERENCES

- R.H. Bundy and R.F. Boyer, *Styrene: Its Polymers, Copolymers and Derivatives*, Reinhold, New York, 1952
- R.T. Cassidy, (Acrylates), in *Engineered Materials Handbook*, Vol. 2, *Engineering Plastics*, ASM International, Metals Park, OH, 1988
- C.A. Harper, Ed., *Handbook of Plastics and Elastomers*, McGraw-Hill, New York, 1975
- J.T. Kroschwitz, Ed., *Encyclopedia of Polymer Science and Engineering*, Vol. 1, Wiley-Interscience, New York, 1985
- R.J. Martino, Ed., *Modern Plastics Encyclopedia*, McGraw-Hill, New York, 1989

- R.J. Martino, (Acrylics), *Modern Plastics*, Vol. 67 (No. 1), 1990, p. 53
- D.C. Miles and J.H. Briston, *Polymer Technology*, Chemical Publishing, New York, 1979
- F.M. McMillan, *The Chain Straighteners*, Macmillan, London, 1979
- H.I. Rowan, (Fluoroplastics), in *Engineered Materials Handbook*, Vol. 2, *Engineering Plastics*, ASM International, Metals Park, OH, 1988
- M.A. Rudner, *Fluorocarbons*, Reinhold, New York, 1958
- H. Sarvetnick, *Polyvinyl Chloride*,"Van Nostrand-Reinhold, New York, 1969
- R.B. Seymour, *Plastics vs. Corrosion*, Wiley-Interscience, New York, 1982
- R.B. Seymour, *Modern Plastics Technology*, Reston Publishing, Reston, VA, 1975
- R.B. Seymour and T. Cheng, Ed., *Advances in Polyethylene*, Plenum Press, New York, 1988
- R.B. Seymour and C.E. Carraher, *Giant Molecules*, Wiley, New York, 1990
- J.W. Summers and E.B. Richardson, (PVC), in *Engineered Materials Handbook*, Vol. 2, *Engineering Plastics*, ASM International, Metals Park, OH, 1988

# CHAPTER 10
# High-Performance Plastics

## INTRODUCTION

According to the *Dictionary of Scientific and Technical Terms*, "Engineering plastics are those which lend themselves to use for engineering design, such as gears and structural members," and according to the *Kirk-Othmer's Encyclopedia of Chemical Technology*, "Engineering plastics are thermoplastics that maintain dimensional stability and most mechanical properties above 100 °C and below 0 °C." The coeditors of *High Performance Polymers: Their Origin and Development* prefer to emphasize "high performance," regardless of whether a polymer is a thermoplastic, thermoset, film, elastomer, or fiber. This broader approach will be used in this chapter, but the emphasis will be on high-performance thermoplastics, discussed in alphabetical order.

## ACETALS

The polymers of formaldehyde (HCHO) are technically termed polyacetals, but the commonly used name "acetal" will be used in this section. Polymers of formaldehyde (polyoxymethylene, POM) have been known from the first time that aqueous solutions of formaldehyde (formalin) were stored at room temperature in the absence of an inhibitor, such as methanol ($CH_3OH$). These deposits do not form at elevated temperatures. Hence, users of large quantities of formalin either added methanol or heated their formaldehyde storage tanks.

Nobel laureate Hermann Staudinger polymerized formaldehyde and noted that the linear polymer was thermally unstable. This "unzipping" was prevented by DuPont chemists, who anionically chain-polymerized formaldehyde from trioxane. They capped the hydroxyl end groups with acetyl groups in 1947. This polymer $((CH_2O))_n$, called Delrin and Tenac, is now produced at an annual rate of more than 30,000 tons. In 1989,

Table 10.1  Properties of Typical Acetal Plastics

| Property | Homo-polymer | Co-polymer | Homo-polymer/ 25% glass | Homo-polymer/ 30% graphite | Co-polymer/ 25% glass |
|---|---|---|---|---|---|
| Melting point ($T_m$), °C | 178 | 170 | 170 | 170 | 170 |
| Heat-deflection temperature at 1.82 MPa, °C | 120 | 105 | 160 | 160 | 160 |
| Maximum resistance to continuous heat, °C | 100 | 105 | 160 | 160 | 160 |
| Coefficient of linear expansion, cm/cm/°C $\times 10^{-5}$ | 10 | 7 | 3 | 5 | 5 |
| Tensile strength, MPa | 67 | 60 | 120 | 52 | 128 |
| % Elongation | 30 | 50 | 2.5 | 1.5 | 3 |
| Flexural strength, MPa | 95 | 90 | 158 | 86 | 193 |
| Compressive strength, MPa | 106 | 110 | 117 | 107 | 117 |
| Notched Izod impact strength, J/m | 128 | 59 | 75 | 40 | 150 |
| Hardness, Rockwell | M92 | M84 | M79 | M80 | M79 |
| Specific gravity | 1.42 | 1.41 | 1.6 | 1.5 | 1.6 |

64,000 tons of acetal plastics were produced in the U.S. In the 1960s, Celanese chemists produced copolymers of formaldehyde with small amounts of ethylene oxide by cationic chain polymerization. This polymer unzipped to produce stable copolymers when heated; called Celon and Kenmetal, they are now produced at an annual rate of more than 30,000 tons in the U.S.

Polyacetals are white, translucent crystalline polymers with a low coefficient of friction. Thus, POM, with 20% polytetrafluoroethylene

Table 10.2  Melting Points and Water Absorptions of Typical Nylons

| Polyamide | $T_m$, °C | H$_2$O absorption, ASTM D570, 24 h | % Amide |
|---|---|---|---|
| Nylon 66 (Zytel) | 265 | 1.15 | 38 |
| Nylon 6 (Grilon) | 226 | 1.16 | 38 |
| Nylon 69 | 226 | 0.5 | 32 |
| Nylon 610 | 216 | 0.5 | 33 |
| Nylon 11 (Rilsan 11) | 184 | 0.4 | 28 |
| Nylon 12 (Rilsan 12, Vestamid) | 179 | 0.3 | 22 |

(PTFE), has been used for bearings. The properties of homoacetals and acetal copolymers are shown in Table 10.1.

Both the homo- and copolymers are available as elastomeric alloys, with impact strengths of the order of 800 to 1000 J/m. Tensile strength has also been increased 20-fold by a 20-fold draw in length.

## NYLONS

Diadic (AB) nylons are produced by the condensation of aliphatic dicarboxylic acids (HOOCRCOOH) and aliphatic diamines (H$_2$NRNH$_2$). Monadic (AA) nylons are produced by ring opening of lactams (CH$_2$(CH$_2$)$_n$NHCO), which can be extruded and injection molded. Amorphous nylons (Trogamid-T) are also available. In addition to nylons produced for fibers, more than 560,000 tons of nylon plastics are used annually throughout the world. In 1989, 270,000 tons of nylon plastics were produced in the U.S.; 70,000 tons of this polyamide (PA) are used as reinforced thermoplastics.

While nylon 66 (62%) and nylon 6 (22%) are the major commercial polyamide engineering plastics, nylon 69, nylon 11, and nylon 12 are also commercially available. As shown in Table 10.2, the melting point decreases and water absorption decreases as the number of methylene groups (CH$_2$) increases or as the percent amide decreases. The properties of nylons are shown in Table 10.3.

Table 10.3 Properties of Typical Nylons

| Property | Nylon 6 | Nylon 6/ 30% glass | Nylon 6/ 30% graphite | Nylon 66 | Nylon 66/ 30% glass | Nylon 66/ 30% graphite | Nylon 66/ 40% clay | Nylon 66/ 50% mica |
|---|---|---|---|---|---|---|---|---|
| Melting point ($T_m$), °C | 226 | 215 | 215 | 265 | 265 | 265 | 265 | 215 |
| Heat-deflection temperature at 1.82 MPa, °C | 78 | 210 | 215 | 75 | 250 | 260 | 190 | 230 |
| Maximum resistance to continuous heat, °C | 65 | 190 | 205 | 100 | 225 | 240 | 150 | 170 |
| Coefficient of linear expansion, cm/cm/°C $\times 10^{-5}$ | 8 | 4 | 5 | 8 | 2 | 2 | 3 | 3 |
| Tensile strength, MPa | 62 | 138 | 205 | 82 | 180 | 227 | 75 | 90 |
| % Elongation | 30 | 5 | 3 | 60 | 4 | 3 | 9 | 9 |
| Flexural strength, MPa | 96 | 275 | 315 | 103 | 275 | 330 | 205 | 400 |
| Compressive strength, MPa | 96 | 150 | 135 | 103 | 180 | 170 | 160 | 150 |
| Notched Izod impact strength, J/m | 55 | 130 | 155 | 55 | 110 | 88 | 50 | 85 |
| Hardness, Rockwell | R119 | M85 | M80 | M85 | M85 | R120 | M80 | M80 |
| Specific gravity | 1.13 | 1.38 | 1.28 | 1.14 | 1.37 | 1.35 | 1.4 | 1.4 |

Table 10.3  Properties of Typical Nylons (continued)

| Property | Nylon 69 | Nylon 610 | Nylon 612 | Nylon 612/ 35% glass | Nylon 11 | Nylon 12 | Aramid |
|---|---|---|---|---|---|---|---|
| Melting point ($T_m$), °C | 205 | 220 | 210 | 210 | 192 | 177 | 275 |
| Heat-deflection temperature at 1.82 MPa, °C | 55 | 60 | 69 | 216 | 150 | 146 | 260 |
| Maximum resistance to continuous heat, °C | 60 | 70 | 75 | 200 | 140 | 135 | 150 |
| Coefficient of linear expansion, cm/cm/°C $\times 10^{-5}$ | 8 | 8 | 8 | 6 | 10 | 8 | 3 |
| Tensile strength, MPa | 58 | 60 | 50 | 145 | 55 | 55 | 120 |
| % Elongation | 80 | 125 | 200 | 4 | 200 | 225 | 5 |
| Flexural strength, MPa | 40 | 40 | 44 | 80 | 40 | 42 | 172 |
| Compressive strength, MPa | 100 | 90 | 90 | 150 | 80 | 80 | 207 |
| Notched Izod impact strength, J/m | 60 | 60 | 60 | 96 | 96 | 110 | 75 |
| Hardness, Rockwell | R111 | R105 | M78 | M93 | R108 | R105 | E90 |
| Specific gravity | 1.09 | 1.08 | 1.08 | 1.35 | 1.04 | 1.01 | 1.2 |

Table 10.4  Properties of Typical PAI Plastics

| Property | PAI | PAI/ 30% glass | PAI/ 30% carbon |
|---|---|---|---|
| Glass transition temperature $(T_g)$, °C | 275 | 275 | 275 |
| Heat-deflection temperature at 1.82 MPa, °C | 275 | 275 | 275 |
| Maximum resistance to continuous heat, °C | 260 | 260 | 260 |
| Coefficient of linear expansion, cm/cm/°C $\times 10^{-5}$ | 3.6 | 1.8 | 2.0 |
| Tensile strength, MPa | 150 | 195 | 205 |
| % Elongation | 13 | 6 | 6 |
| Flexural strength, MPa | 200 | 315 | 315 |
| Compressive strength, MPa | 258 | 300 | 300 |
| Notched Izod impact strength, J/m | 135 | 105 | 45 |
| Hardness, Rockwell | E78 | E94 | E94 |
| Specific gravity | 1.39 | 1.57 | 1.41 |

## POLYAMIDE-IMIDES (PAI)

Polyamide-imides (Torlon) are produced by the condensation of trimellitic trichloride $[(C_6H_3)(COCl)_3]$ with aromatic amines, such as methylenedianiline $(H_2NC_6H_4)_2CH_2)$. This amber-gray amorphous resin is more resistant to moisture than crystalline nylon. PAI's are useful over a temperature range of −160 to 230 °C. These engineering plastics have a limiting oxygen index (LOI) of 43% and a UL 94 rating of V-0. They emit very little smoke when burned, and thus exceed FAA regulations for materials used in commercial aircraft interiors. PAI's are used in combustion engines as piston skirts, push rods, turbocharger rotors, and stators. As shown by the data in Table 10.4, unfilled PAI is one of the strongest plastics available. PAI has been used in many applications where resistance to elevated temperatures is essential. The most dramatic

Table 10.5  Properties of a Typical Polyarylate

| Property | |
| --- | --- |
| Heat-deflection temperature at 1.82 MPa, °C | 174 |
| Maximum resistance to continuous heat, °C | 150 |
| Coefficient of linear expansion, cm/cm/°C $\times 10^{-5}$ | 6.5 |
| Tensile strength, MPa | 68 |
| % Elongation | 50 |
| Flexural strength, MPa | 74 |
| Compressive strength, MPa | 93 |
| Notched Izod impact strength, J/m | 210 |
| Hardness, Rockwell | R125 |
| Specific gravity | 1.2 |

application is in the Polimoter combustion engine. Annual production of PAI is 90 tons.

## POLYARYLATES

Polyarylates are condensation products of aromatic dicarboxylic acyl chlorides, such as terephthaloyl chloride $(C_6H_4(COCl)_2$, and diphenols, such as bisphenol A $(HOC_6H_4)_2C(CH_3)_2$. These clear, amorphous, heat-resistant polymers are marketed under the trade names of Ardel and Durel. Liquid crystalline arylates (Ekcel, Ekonol) are also being used for high-temperature applications.

These transparent, aromatic polyesters have been used for automotive lenses and sodium lamps. However, polyarylates undergo a photo-Fries rearrangement and darken when exposed to light for long periods of time. Polyarylates are used for electronic components and firefighter helmets. The properties of a typical polyarylate are summarized in Table 10.5.

## POLYCARBONATES (PC)

Polycarbonates are tough, amorphous, transparent polymers that were developed independently in the 1950s by Schnell of Bayer AG and Fox of General Electric. They are produced by the condensation of phosgene ($COCl_2$) and bisphenol A (2,2-*bis*(hydroxyphenyl) propane) and marketed under the trade names of Makrolon, Merlon, Calibre, Lexan, and Panlite. Like other amorphous polymers, polycarbonates are attacked by chlorinated hydrocarbon solvents and stress crack in contact with ketones, such as acetone and methyl ethyl ketone.

Polycarbonate resins are injection moldable, extrudable, and blow moldable. PC sheet is used for glazing, signs, and furniture. Annual worldwide sales of PC are 282,000 tons; 20,000 tons of this polymer are used as reinforced thermoplastics. Polyester carbonate resins are produced by adding phthaloyl dichloride with phosgene to the reactants with bisphenol A. The polyester carbonates are resistant to higher temperatures and more resistant to hydrolysis than PC. The properties of a typical polycarbonate sheet are shown in Table 10.6.

## POLYESTERS

In contrast to the arylates, which are aromatic polyesters, and polycarbonates, which are esters of carbonic acid, the simple name "polyester" is used to describe aromatic/aliphatic polymers produced by the condensation of terephthalic and ethylene glycol. These polyethylene terephthalates (PET) are marketed under the trade names of Arnite, Crastin, Petlon, Petra, Rynite, and Vestodur. PET, of course, is also widely used as a synthetic fiber.

Polybutylene terephthalate (PBT) is marketed under the trade names of Arnite, Celanex, Duranex, Gafite, Gaftuf, Pibiter, Valox, Techster, and Ultradur. The nonfiber applications of PET include blow-molded bottles and films. Both PET and PBT are used for injection molding and extrusion. In 1989, 995,000 tons of PET and PBT engineering plastics were produced in the U.S.; 50,000 tons were used as reinforced thermoplastics.

A blend of PBT and elastomers is sold by G.E. under the trade name Lomod. The Xenoy blend of PBT and PC is called Zenoy by G.E. and Makroblend by Mobay. Polybutylene terephthalate/polyphenylene oxide (PBT/PPO) alloys are sold by G.E. under the trade name Gemax. It is

Table 10.6 Properties of Typical Polycarbonate Sheet

| Property | PC | PC/ 10% gloss | PC/ 30% gloss | PC/ 30% graphite | PC/ 40% graphite | Polyester carbonate |
|---|---|---|---|---|---|---|
| Glass transition temperature ($T_g$), °C | 150 | 150 | 150 | 150 | 150 | 160 |
| Heat-deflection temperature at 1.82 MPa, °C | 130 | 142 | 144 | 144 | 146 | 150 |
| Maximum resistance to continuous heat, °C | 125 | 130 | 130 | 130 | 130 | 135 |
| Coefficient of linear expansion, cm/cm/°C $\times 10^{-5}$ | 7 | 2 | 1 | 1 | 1 | 8 |
| Tensile strength, MPa | 65 | 65 | 135 | 165 | 165 | 73 |
| % Elongation | 110 | 6 | 3 | 3 | 2 | 90 |
| Flexural strength, MPa | 93 | 105 | 155 | 93 | 240 | 240 |
| Notched Izod impact strength, J/m | 130 | 110 | 90 | 100 | 90 | 300 |
| Hardness, Rockwell | M70 | M75 | M92 | R118 | R119 | M85 |
| Specific gravity | 1.2 | 1.28 | 1.4 | 1.35 | 1.35 | 1.2 |

Table 10.7 Properties of Typical Polyester Resins

| Property | PET | PET/ 30% glass | PET/ 45% glass | PET/ 30% graphite | PBT | PBT/ 30% glass | PBT/ 45% glass | PBT/ 30% graphite |
|---|---|---|---|---|---|---|---|---|
| Melting point ($T_m$), °C | 255 | 255 | 255 | 255 | 245 | 245 | 245 | 245 |
| Heat-deflection temperature at 1.82 MPa, °C | 220 | 220 | 225 | 225 | 85 | 210 | 210 | 215 |
| Maximum resistance to continuous heat, °C | 200 | 210 | 210 | 210 | 80 | 200 | 200 | 205 |
| Coefficient of linear expansion, cm/cm/°C $\times 10^{-5}$ | 6.5 | 3 | 2 | 3 | 6 | 2.5 | 2 | 3 |
| Tensile strength, MPa | 58 | 150 | 165 | 175 | 50 | 220 | 90 | 155 |
| % Elongation | 100 | 4 | 3 | 115 | 100 | 3 | 3 | 2 |
| Flexural strength, MPa | 110 | 225 | 175 | 260 | 100 | 175 | 140 | 215 |
| Compressive strength, MPa | 90 | 170 | 155 | 105 | 98 | 145 | 105 | 100 |
| Notched Izod impact strength, J/m | 35 | 80 | 75 | 75 | 40 | 50 | 50 | 70 |
| Hardness, Rockwell | M97 | M100 | R118 | R125 | M72 | M90 | M80 | R120 |
| Specific gravity | 1.35 | 1.6 | 1.6 | 1.4 | 1.34 | 1.5 | 1.7 | 1.41 |

Table 10.8  Properties of Typical PEEK Plastics

| Property | PEEK | PEEK/ 30% glass | PEEK/ 30% graphite |
|---|---|---|---|
| Melting point ($T_m$), °C | 334 | 334 | 334 |
| Heat-deflection temperature at 1.82 MPa, °C | 165 | 282 | 282 |
| Maximum resistance to continuous heat, °C | 150 | 270 | 270 |
| Coefficient of linear expansion, cm/cm/°C $\times 10^{-5}$ | 5.5 | 2.1 | 1.5 |
| Tensile strength, MPa | 100 | 162 | 173 |
| % Elongation | 40 | 2 | 2 |
| Flexural strength, MPa | 110 | 255 | 313 |
| Notched Izod impact strength, J/m | 150 | 110 | 70 |
| Hardness, Rockwell | R123 | R123 | R123 |
| Specific gravity | 1.32 | 1.44 | 1.32 |

customary to use PBT/10% nylon or PET/15% high-density polyethylene (HDPE) for fiberglass-reinforced PBT. The surface gloss of PET is improved by blending with PBT. The properties of polyester engineering resins are shown in Table 10.7.

## POLYETHER ETHER KETONES (PEEK)

By recognizing structure-property relationships in 1970, Rose designed a polyether ether ketone which, because of the presence of a carbonyl stiffening group, was temperature resistant and strong, yet had some flexibility as a result of the presence of flexibilizing ether groups in the polymer chain. This partially crystalline polymer also possesses a high degree of oxidation stability and radiation, solvent, and flame resistance. It has an LOI of 34 to 35 and a UL rating of V-0. PEEK, which is used in many military and nuclear energy applications, is marketed by ICI under

Table 10.9  Properties of Typical PAEK Plastics

| Property | PAEK | PAEK/ 30% glass |
|---|---|---|
| Melting point ($T_m$), °C | 340 | ... |
| Heat-deflection temperature at 1.82 MPa, °C | 160 | 325 |
| Coefficient of linear expansion, cm/cm/°C $\times 10^{-5}$ | 4.7 | 2.2 |
| Tensile strength, MPa | 93 | 163 |
| % Elongation | 50 | 2 |
| Flexural modulus, GPa | 3.6 | 10 |
| Notched Izod impact strength, J/m | 85 | 96 |
| Dielectric constant | 3.2 | 3.7 |
| Oxygen index, % | 34 | 40.5 |

the trade name of Victrex. Less than 500 tons of PEEK are produced annually, but this volume is growing at the rate of about 10%.

The properties of PEEK are shown in Table 10.8. PEEK is produced by the condensation of 4,4′-difluorodiphenyl ketone (($FC_6H_4)_2$ C=O) with the potassium salt of hydroquinone ($KOC_6H_4OK$).

Another commercial polyarylether ketone (PAEK) is sold under the trade name Kodel by Amoco. PAEK has excellent thermal stability over a range of 65 °C and hence is readily processed at temperatures below its melting point of 340 °C. It has a UL 94 rating of V-0. The properties of typical PAEK's are shown in Table 10.9.

## POLYETHERIMIDES (PEI)

Polyimides, which are used both as thermosets and thermoplastics, lack flexibility, but this deficiency can be overcome by making polyamide-imides (discussed previously) or by making polyetherimides. These amorphous, transparent resins are produced by the condensation of a diamine and an aromatic dianhydride.

PEI, which is marketed under the trade name of Ultem by G.E., is resistant to flame, radiation, and most solvents, but is soluble in some

Table 10.10 Properties of Typical PEI Plastics

| Property | PEI | PEI/ 10% glass | PEI/ 20% glass | PEI/ 30% glass | PEI/ 30% graphite |
|---|---|---|---|---|---|
| Glass transition temperature, °C | 216 | 216 | 216 | 216 | 216 |
| Heat-deflection temperature at 1.82 MPa, °C | 195 | 200 | 205 | 210 | 210 |
| Maximum resistance to continuous heat, °C | 165 | 170 | 175 | 180 | 180 |
| Coefficient of linear expansion, cm/cm/°C $\times 10^{-5}$ | 5 | 4 | 3 | 2 | 2 |
| Tensile strength, MPa | 105 | 114 | 138 | 169 | 216 |
| % Elongation | 7 | 5 | 4 | 3 | 2 |
| Flexural strength, MPa | 144 | 193 | 205 | 225 | 283 |
| Compressive strength, MPa | 140 | 155 | 162 | 175 | 220 |
| Notched Izod impact strength, J/m | 55 | 60 | 85 | 110 | 75 |
| Hardness, Rockwell | M110 | M116 | M120 | M125 | M127 |
| Specific gravity | 1.3 | 1.35 | 1.45 | 1.5 | 1.4 |

chlorinated aliphatic solvents. PEI has an oxygen index rating of 47%, a UL 94 rating of V-0, and a dielectric constant that is essentially unchanged with increases in frequencies up to $10^9$ Hz. This engineering polymer has been used in automotive and aerospace applications and in wire insulation. The properties of a typical PEI resin are shown in Table 10.10.

## THERMOPLASTIC POLYIMIDES (PI)

Thermoplastic polyimides can be produced by the condensation of an aromatic diisocyanate. Filaments of ordered PI resins (Vespel) can be spun from a methane sulfonic acid solution and used as reinforcing fila-

| Table 10.11  Properties of Typical Polyimides | | |
|---|---|---|
| **Property** | **PI** | **PI/ 40% graphite** |
| Glass transition temperature, °C | 330 | 365 |
| Heat-deflection temperature at 1.82 MPa, °C | 315 | 360 |
| Maximum resistance to continuous heat, °C | 290 | 810 |
| Coefficient of linear expansion, cm/cm/°C $\times\ 10^{-5}$ | 5 | 4 |
| Tensile strength, MPa | 96 | 44 |
| % Elongation | 9 | 3 |
| Flexural strength, MPa | 165 | 145 |
| Compressive strength, MPa | 240 | 125 |
| Notched Izod impact strength, J/m | 83 | 38 |
| Hardness, Rockwell | E70 | E27 |
| Specific gravity | 1.4 | 1.65 |

ments in epoxy resins. Because of their outstanding thermal stability, based on the high electron resonance structure of the backbone, many different types of polyimides have been synthesized by both condensation and addition techniques.

Some PI prepolymers are processed as polyamic acid precursors to intractable polymers. Skybond and Pyralin PI's are cured by closing an imide ring in the precursor. Other PI's, such as PMR-15, have ester end caps, while Thermid has acetylene end caps through which curing takes place. However, fully imidized thermoplastic PI's are available from Ciba Geigy (XU 218), Upjohn (2080), and NASA (Larc-TPI). Polyimides are used in automotive and aerospace applications. About 400 tons of PI are produced annually in the U.S. The properties of typical polyimides are shown in Table 10.11.

| Table 10.12  Properties of PBI | |
|---|---|
| **Property** | |
| Glass transition temperature $(T_g)$, °C | 425 |
| Heat-deflection temperature at 1.82 MPa, °C | 435 |
| Coefficient of linear expansion, cm/cm/°C $\times 10^{-5}$ | 2.3 |
| Tensile strength, MPa | 160 |
| Tensile modulus, GPa | 5.8 |
| % Elongation | 3 |
| Flexural strength, MPa | 220 |
| Flexural modulus, GPa | 6.5 |
| Compressive strength, MPa | 390 |
| Compressive modulus, GPa | 5.8 |
| Notched Izod impact strength, J/m | 30 |
| Hardness, Rockwell | 110K |
| Dielectric constant | 3.3 |
| Limiting oxygen index, % | 58 |

## POLYBENZIMIDAZOLES (PBI)

The principal polybenzimidazole is poly(2,2′-*m*-phenylene)-5,5′dibenzimidazole, which is produced by the hot melt condensation reaction of 3,3′-diaminobenzidine and diphenylisophthalate. PBI is one of the few organic plastics that is able to maintain a relatively high percentage of its load-bearing properties for as much as 5 min at 260 °C. About 25% of the compressive strength of PBI is retained after a dwell time of 30 min at 315 °C. There is a linear decrease of the compressive strength of 420 MPa at −40 °C to that of 140 MPa at 315 °C. The properties of Celazole PBI are shown in Table 10.12.

Table 10.13  Properties of Typical PPS Resins

| Property | PPS | PPS/ 40% glass | PPS/ 40% graphite |
|---|---|---|---|
| Melting point ($T_m$), °C | 290 | 290 | 280 |
| Heat-deflection temperature at 1.82 MPa, °C | 133 | 260 | 260 |
| Maximum resistance to continuous heat, °C | 120 | 240 | 240 |
| Coefficient of linear expansion, cm/cm/°C $\times 10^{-5}$ | 5 | 2 | 1 |
| Tensile strength, MPa | 65 | 135 | 160 |
| % Elongation | 2 | 2 | 1.5 |
| Flexural strength, MPa | 95 | 185 | 210 |
| Compressive strength, MPa | 95 | 160 | 180 |
| Notched Izod impact strength(a), J/m | 25 | 80 | 55 |
| Hardness, Rockwell | R123 | R123 | R123 |
| Specific gravity | 1.3 | 1.65 | 1.45 |
| Dielectric constant | 3.9 | 3.9 | ... |

(a) Impact values are much higher before annealing.

## POLYPHENYLENE ETHER (PPO/PPE)

More than 70,000 tons of PPO are consumed annually by the American plastics industry. Both PPO and Noryl are trademarks of G.E. for blends of polyphenylene ether and polystyrene. These products are discussed in Chapter 6.

## POLYPHENYLENE SULFIDE (PPS)

Polyphenylene sulfide (Ryton, Tedor, Supec, Fortran) has been produced by the Wurtz-Fittig condensation of $p$-dichlorobenzene ($ClC_6H_4Cl$) and sodium sulfide ($Na_2S$) under the trade name of Ryton by Phillips Petroleum Company since 1970. This brown-black, crystalline, heat-resistant and chemical-resistant polymer can be injection molded and extruded to produce automobile, aerospace, and chemical processing components. PPS fiber, films, and composites are commercially available. Post-annealing after processing is recommended to ensure dimensional stability of crystalline PPS. The properties of typical PPS products are shown in Table 10.13.

## POLYSULFONES (PSO, PES)

The first commercial polyarylsulfone resin (Udel) was introduced by Union Carbide in 1966. Since then, BASF, ICI, and 3M have introduced other polysulfones under the trade names of Radel, Victrex, and Astrel. These engineering polymers contain sulfonyl ($SO_2$) stiffening groups in the polymer chain and are produced by Friedel-Crafts condensation ($AlCl_3$) of reactants with sulfonyl groups. These reactants are the potassium salt of bisphenol A (Udel and Radel), the sodium salt of 4,4'-dihydroxydiphenyl sulfone (Victrex), and 4,4'-dichlorosulfonyldiphenyl ether (Astrel).

These amorphous yellow-amber transparent resins, which have a high degree of thermoxidative and hydrolytic stability and low resistance to sunlight, are used for components of electrical, electronic, and medical devices. These engineering plastics are commercially available from Amoco and ICI under the trade names of Radel (polyarylsulfone), Victrex (polyethersulfone, PES) and Udel (polysulfone, PSO). All of these plastics are characterized by good retention of physical and electrical properties over a temperature range of −60 to 200 °C. These plastics have an LOI value of about 40% and meet UL 94 V-0 flammability specifications. These plastics can be plated by an electroless nickel or copper process. About 10,000 tons of these high-performance plastics are used annually in the U.S. The properties of polysulfones are summarized in Table 10.14.

## LIQUID CRYSTAL POLYMERS (LCP)

Several aromatic polyesters, described earlier in this chapter, meet the specifications for anisotropic liquid crystals. These low-viscosity, injection-moldable and extrudable, heat-resistant aromatic polyesters undergo parallel ordering in the molten state, which is evident by the presence of tightly packed fibrous chains in the fabricated moldings and extrudates.

Table 10.14  Properties of Typical Polysulfones

| Property | PSO (Udel) | PSO/ 30% glass | PSO/ 30% graphite | PES (Victrex) | PES/ 20% glass | PES/ 30% graphite | Modified PSO | Modified PSO/ 30% glass |
|---|---|---|---|---|---|---|---|---|
| Heat-deflection temperature at 1.82 MPa, °C | 190 | 198 | 190 | 200 | 210 | 210 | 150 | 150 |
| Maximum resistance to continuous heat, °C | 170 | 175 | 175 | 185 | 200 | 200 | 150 | 150 |
| Coefficient of linear expansion, cm/cm/°C $\times 10^{-5}$ | 6 | 2.5 | 0.5 | 5.5 | 2 | 1 | 4 | 5 |
| Tensile strength, MPa | 70 | 100 | 160 | 138 | 127 | 190 | 43 | 115 |
| % Elongation | 5 | 115 | 115 | 85 | 2 | 115 | 50 | 2 |
| Flexural strength, MPa | 106 | 200 | 215 | 120 | 175 | 250 | 85 | 175 |
| Compressive strength, MPa | 176 | 95 | 175 | 95 | 150 | 150 | 125 | 150 |
| Notched Izod impact strength, J/m | 64 | 58 | 64 | 110 | 75 | 75 | 150 | 75 |
| Hardness, Rockwell | M69 | M95 | M80 | M88 | M98 | R123 | R117 | M80 |
| Specific gravity | 1.25 | 1.5 | 1.36 | 1.4 | 1.5 | 1.5 | 1.35 | 1.5 |

Table 10.15  Properties of Typical Liquid Crystal Polymers

| Property | LCP | LCP/ 50% talc |
|---|---|---|
| Glass transition temperature $(T_g)$, °C | 400 | 525 |
| Heat-deflection temperature at 1.82 MPa, °C | 350 | 525 |
| Maximum resistance to continuous heat, °C | 250 | 250 |
| Tensile strength, MPa | 135 | 80 |
| % Elongation | 4 | 3 |
| Flexural strength, MPa | 13 | 115 |
| Compressive strength, MPa | 42 | 42 |
| Notched Izod impact strength, J/m | 165 | 75 |
| Hardness, Rockwell | R66 | R76 |
| Specific gravity | 1.35 | 1.85 |

These nematic self-reinforcing polymers, which can withstand intermittent temperatures as high as 315 °C, have a UL rating of 240 °C and a UL 94 rating of V-0. These wholly aromatic engineering polymers are marketed under the trade names of Xydar, Calendar, Vectra, and Polyester X7G by Dartco, Celanese, and Eastman at an annual rate of 10,000 tons. The properties of typical LCP's are shown in Table 10.15.

## REINFORCED THERMOSETS

High-performance reinforced thermoplastics have displaced metals, ceramics, and reinforced thermosets in many applications. Nevertheless, the reinforced thermosets, commonly called reinforced plastics, are used at an annual rate of more than 250,000 tons in the U.S. These composites are described in Chapter 11.

# REFERENCES

- J. Bevilaqua, (Polyetherimide), *Modern Plastics*, Vol. 66 (No. 11), 1989, p. 41
- G.V. Cekis, (Polyamide-imide), *Modern Plastics*, Vol. 66 (No. 11), 1989, p. 32
- S.R. Dunkle and B.D. Dean, (Polyarylates), *Modern Plastics*, Vol. 67 (No. 11), 1990, p. 34
- J.N. Epel, J.M. Margolis, S. Newman, and R.B. Seymour, Ed., *Engineered Materials Handbook*, Vol. 2, *Engineering Plastics*, ASM International, Metals Park, OH, 1988
- D.P. Garner and G.A. Stahl, Ed., *The Effects of Hostile Environments on Coatings and Plastics*, ACS Symposium Series 229, American Chemical Society, Washington, DC, 1983
- G.M. Grayson, Ed. *Encyclopedia of Composite Materials and Components*, John Wiley & Sons, New York, 1983
- W. Hoven-Nievelstein, (Acetals), *Modern Plastics*, Vol. 66 (No. 11), 1989, p. 19
- P.J. Huspeni, R. Layton, and M. Matzner, (LCP), *Modern Plastics*, Vol. 66 (No. 11), 1989, p. 42
- D.J. Kemmish, *Modern Plastics*, Vol. 66 (No. 11), 1989, p. 26
- M.I. Kohan and R.L. Ward, (Nylons), *Modern Plastics*, Vol. 66 (No. 11), 1989, p. 30
- G.R. Kriek, O.C. Mitschke, and J. Sami, (PET/PBI), *Modern Plastics*, Vol. 43, 1989, p. 45
- J.I. Kroschwitz, Ed., *Encyclopedia of Polymer Science and Engineering*, Vol. 6, Wiley-Interscience, New York, 1986
- G. Lubin, Ed. *Handbook of Composites*, Van Nostrand Reinhold, New York, 1982
- J.M. Margolis, Ed., *Engineering Thermoplastics*, Marcel Dekker, New York, 1985
- R.J. Martino, Ed., *Modern Plastics Encyclopedia*, McGraw-Hill, New York, 1989
- L.A. McKenna, (Polysulfone), *Modern Plastics*, Vol. 66 (No. 11), 1989, p. 118
- D.C. Miles and J.H. Briston, *Polymer Technology*, Chemical Publishing, New York, 1979

- H.R. Penton, (Polyimide), *Modern Plastics*, Vol. 66 (No. 11), 1989, p. 87
- T.S. Renegar, (Polyphenylene sulfide), *Modern Plastics*, Vol. 66 (No. 11), 1989, p. 97
- H. Saechling, *International Plastics Handbook*, Hanser Verlag, Munich, 1983
- R.B. Seymour and G.S. Kirshenbaum, Ed., *High Performance Polymers: Their Origin and Development*, Elsevier Science, New York, 1986
- R.B. Seymour, *Polymers for Engineering Applications*, American Society for Metals, Metals Park, OH, 1987
- R.B. Seymour, *Plastics vs. Corrosives*, Wiley-Interscience, New York, 1982
- R.B. Seymour and C.E. Carraher, *Polymer Chemistry: An Introduction*, Marcel Dekker, New York, 1988
- R.B. Seymour, *Engineering Polymers Source Book*, McGraw Hill, New York, 1989
- R.W. Tess and G.W. Poehlein, Ed., *Applied Polymer Science*, ACS Symposium Series 285, American Chemical Society, Washington, DC, 1985
- W.V. Titow and B.J. Lanham, *Reinforced Thermoplastics*, John Wiley & Sons, New York, 1975

# CHAPTER 11

# Reinforced Thermosets

## INTRODUCTION

Fibers used for reinforcing polymers were described in Chapter 4. Thermoplastic composites with these additives were discussed in Chapter 10. Thermoset composites will be discussed in this chapter.

According to IAL Consultants (London), annual sales of reinforced plastics in Europe exceed 600,000 tons and are valued at about $1.25 billion. According to the Society of the Plastics Industry, more than 1 million tons of plastic composites were produced in the U.S. in 1988; the automobile and construction industries continue to be the first and second largest consumers.

## ENERGY REQUIREMENTS

The advantage of using composites that weigh only 20% as much as steel is obvious and essential for economical air, land, and marine transportation. Likewise, the energy required for the production of aluminum sheet is three times that required for producing a composite sheet by pultrusion. Of course, die-cast aluminum uses half the energy required for aluminum sheet, and spray-up and filament-wound components require 40% more energy than pultrusion.

It is important to note that cost savings are not limited to energy costs for fabrication. Most metal fabrication involves several labor-intensive steps, while most reinforced plastics require fewer parts to produce complex units via less labor-intensive automated fabrication techniques.

## EFFECT OF FIBER ORIENTATION
## ON PROPERTIES

The optimum unidirectional strength of components is observed with continuous filaments. However, as shown by the data in Table 11.1, the

Table 11.1  Properties of Polyester Plastics Reinforced by
Continuous and Chopped Fiberglass

| Continuous filament, % | Chopped glass, % | Tensile strength, MPa | Flexural strength, MPa | Transverse tensile strength, MPa | Transverse flexural strength, MPa |
|---|---|---|---|---|---|
| 75 | 0 | 690 | 1200 | 24 | 35 |
| 65 | 10 | 660 | 1135 | 27 | 90 |
| 45 | 20 | 570 | 980 | 60 | 155 |
| 25 | 50 | 500 | 810 | 95 | 200 |
| 15 | 60 | 410 | 680 | 125 | 260 |

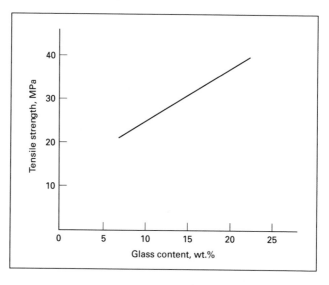

*Fig. 11.1  Tensile strength versus glass content*

transverse strength is improved by replacing some of the 75% continuous glass filament reinforcements with chopped fiberglass. As shown in Fig. 11.1, the tensile strength of fiber-reinforced plastics (FRP) is proportional to the content of fiberglass.

Table 11.2 Properties of Epoxy Resin Composites With Different Reinforcing Fibers

| Property | E-glass | S-glass | Aramid (Kevlar 49) | Graphite | Boron |
|---|---|---|---|---|---|
| Thermal conductivity, W/m · K | 0.9 | 1.1 | 0.9 | 5 | 1 |
| Coefficient of linear expansion, cm/cm/°C $\times 10^{-5}$ | 1.2 | 1.1 | 1 | 2 | 1 |
| Tensile strength, MPa | 450 | 700 | 800 | 700 | 1600 |
| Elastic modulus, GPa | 24 | 30 | 33 | 60 | 207 |
| Fracture toughness, MPa · m$^{1/2}$ | 2.1 | 2.0 | 1.4 | 1.6 | 2.1 |

## COMPARISON OF FIBER TYPE
## ON PROPERTIES OF COMPOSITES

Since the strength of the resin matrix is only 1 or 2% that of the reinforcing fiber, its principal function is as a binder, which permits transfer of the applied load to the fiber. Accordingly, some purists suggest that the term "reinforced plastics" be replaced by the term "bonded fiber composites." In any case, as shown in Table 11.2, the combination of resin and reinforcing agent provides a composite with specific physical properties that exceed those of most other materials of construction. Specific properties are those based on volume rather than mass.

As stated in Chapter 4, the physical properties of composites are dependent, to a large extent, on the type of resin and reinforcing fiber used. As shown in Table 11.2, most of the properties of epoxy resin composites are improved as one goes from E-glass to S-glass, to aramid, to graphite, and to boron filament reinforcements.

E-glass fiber composites are characterized by higher impact energy than carbon or boron composites. Young and Brantemar correlated impact energy and the interlaminar shear strength of laminated composites and showed that fiber failure was the principal failure at high interlaminar shear strength. Since epoxy resins have better adhesion than polyesters to

Table 11.3  Typical CTE Values for Materials of Construction(a)

| Material | Longitudinal | Transverse | Quasi-isotropic |
|---|---|---|---|
| S-glass/epoxy | 6.3 | 19.8 | 10.8 |
| Kevlar 49/epoxy | –3.6 | 54 | –0.9 to +0.9 |
| Graphite/epoxy | 0.9 | 27 | 0 to 0.9 |
| Boron/epoxy | 4.5 | 14.4 | 3.6 to 5.4 |
| Aluminum | ... | 21.6 to 25.2 | ... |
| Steel | ... | 10.8 to 18 | ... |
| Epoxy | ... | 54 to 90 | ... |

(a) CTE = cm/cm/°C $\times$ $10^{-6}$ (laminates = 60% of volume). Data adapted from Wit. Freeman and G.C. Kuboler, *ASTM STP 546*, 1974, p. 435

Table 11.4  Effect of Reinforcement on Thermal Conductivity (W/m · K) of Epoxy Resin Laminates

| Reinforcing fiber | Longitudinal | Transverse | Quasi-isotropic |
|---|---|---|---|
| None | ... | 0.35 | ... |
| S-glass | 3.5 | 0.35 | 0.35 |
| Aramid | 1.7 | 0.7 | 0.17 |
| High modular carbon | 5.5 | 1 | 15 |
| Ultrahigh modular carbon | 125 | 0.04 | 27 |
| Boron | 1.7 | 1.0 | 1.4 |

E-glass, epoxy glass composites have higher impact energy and fail through a combinations of fiber failure and delamination.

The impact resistance of composites may be regulated by using hybrid combinations of fibers such as E-glass and carbon fiber. The coefficient of thermal expansion (CTE), which is less than that of unreinforced plastics, can also be regulated by the use of hybrid fibers. Aramid and carbon fiber composites have negative values for this coefficient, and glass or boron fibers have positive expansion values in the longitudual direction.

This anisotropicity can be reduced; i.e., quasi-isotropic laminates can be produced by the use of disoriented staple fibers. The coefficients of expansion ($10^{-6}/°C$) for specific epoxy resin laminates are shown in Table 11.3.

With the exception of carbon, most reinforcing fibers have low thermal conductivity values which are directionally dependent, as shown by the data in Table 11.4. The relative electrical conductivity values of these laminates parallels those shown for thermal conductivity. Most mechanical properties of plastics composites, like those of the resin matrix, are lowered when the temperature is increased up to the resin deterioration temperature. This decrease is greater under high humidity conditions.

## COMPARISON OF RESIN TYPE ON PROPERTIES OF COMPOSITES

As shown in Table 11.5, the principal effect of the resin matrix is on the heat-deflection temperature of selected fiberglass-reinforced composites.

## COMPARISON OF FABRICATION PROCEDURE ON PROPERTIES OF POLYESTER COMPOSITES

As shown by the data in Table 11.6, the properties of anisotropic filament-wound and pultruded composites, measured along the fiber axis, are much higher than those of the more isotropic composites.

## EFFECT OF TEMPERATURE AND BOILING WATER ON PHYSICAL PROPERTIES OF COMPOSITES

As shown in Fig. 11.2, the tensile strength of an SMC composite with 25% fiberglass decreases dramatically when the temperature is increased from −40 to 149 °C. The effect of temperature on the linear coefficient of expansion of an SMC composite with 20% fiberglass is shown in Fig. 11.3.

As shown in Fig. 11.4, the flexural strength of an SMC polyester composite containing 20% fiberglass decreases after 15 days in boiling water. The bisphenol A polyester is the most resistant to boiling water and the ortho-phthalate polyester is the least resistant. The properties of all of these polyesters are improved when the fiberglass is coated with properly selected coupling agents.

Table 11.5 Properties of Fiberglass Composites With Different Thermosets

| Property | Diallyl phthalate | Epoxy | Phenolic | Polyester | Polyamide |
|---|---|---|---|---|---|
| Heat-deflection temperature at 1.82 MPa, °C | 225 | 150 | 200 | 200 | 350 |
| Maximum resistance to continuous heat, °C | 210 | 140 | 175 | 160 | 310 |
| Coefficient of linear expansion, cm/cm/°C $\times 10^{-5}$ | 3 | 2 | 2 | 2.5 | 1.3 |
| Tensile strength, MPa | 50 | 83 | 41 | 70 | 44 |
| % Elongation | 4 | 4 | 1.5 | 1 | 1 |
| Flexural strength, MPa | 90 | 103 | 172 | 172 | 145 |
| Compressive strength, MPa | 205 | 100 | 200 | 200 | 300 |
| Notched Izod impact strength, J/m | 50 | 25 | 175 | 200 | 300 |
| Hardness, Rockwell | E84 | M105 | M110 | M50 | M118 |
| Specific gravity | 1.8 | 1.9 | 1.5 | 2 | 1.6 |

Table 11.6 Properties of Fiberglass-Reinforced Polyester Composites Made by Different Fabrication Techniques

| Property | SMC | BMC | Preform mat | Cold-press molding | Spray-up | Filament-wound (epoxy) | Pultruded |
|---|---|---|---|---|---|---|---|
| Glass content, % | 22 | 25 | 30 | 25 | 40 | 55 | 60 |
| Heat-deflection temperature at 1.82 MPa, °C | 225 | 225 | 205 | 190 | 190 | 190 | 175 |
| Maximum resistance to continuous heat, °C | 180 | 175 | 185 | 180 | 185 | 150 | 200 |
| Coefficient of linear expansion, cm/cm/°C $\times 10^{-5}$ | 1.0 | 1.0 | 1.4 | 1.4 | 1.6 | 4 | 5 |
| Tensile strength, MPa | 90 | 48 | 110 | 110 | 95 | 120 | 80 |
| % Elongation | 1 | 0.5 | 1.5 | 1.5 | 1.0 | 2.0 | 2.0 |
| Flexural strength, MPa | 165 | 100 | 220 | 190 | 150 | 1250 | 1000 |
| Compressive strength, MPa | 80 | 30 | 150 | 125 | 135 | 400 | 340 |
| Notched Izod impact strength, J/m | 640 | 240 | 800 | 560 | 425 | 2660 | 2750 |
| Hardness, Rockwell | H75 | H95 | H70 | H70 | H70 | M110 | H96 |
| Specific gravity | 1.9 | 1.9 | 1.9 | 1.6 | 1.5 | 1.9 | 1.8 |

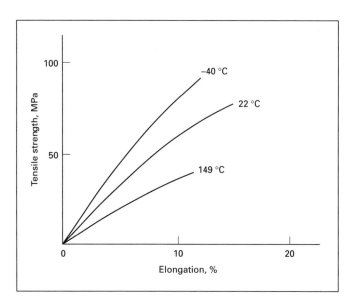

*Fig. 11.2 Properties of SMC composite versus temperature*

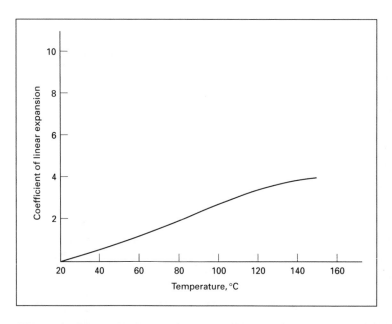

*Fig. 11.3 Effect of temperature on coefficient of linear expansion of SMC (cm/cm/°C × 10⁻⁵)*

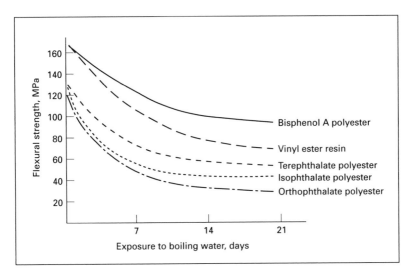

Fig. 11.4  Effect of boiling water on SMC composites

# AIRCRAFT APPLICATIONS

The 'round-the-world flight, without refueling, of the *Challenger* in 1987 demonstrated the distinct advantage of plastic composites in aircraft structures. This advantage is also demonstrated by the Bell Textron V22 Osprey helicopter, which uses graphite-epoxy composites in its rotor and is 25% lighter than helicopters made from conventional materials. The yacht that won the America Cup in 1987 was also constructed from plastic composites.

The replacement of metals by polymer composites in helicopters has resulted in a more readily assembled unit with 9000 fewer components, weight reduction, improved reliability, and increased fuel economy. The volume of reinforcing fibers used in aircraft in 1989 was as follows: 750,000 tons of fiberglass, 1600 tons of graphite fibers, and 800 tons of aramid fibers. The amphibious *Sea Wind* was made from reinforced vinyl ester resin.

The 10-passenger Lear Fan Jet aircraft made from graphite and aramid-reinforced epoxy resin (CFRP) received much publicity because of financial and design problems, light weight, and fuel economy. CFRP is also used for the Beech Starship C-EP and filament-wound rocket booster engines. McDonnell Douglas is using graphite-reinforced epoxy composites for primary fuselage wing areas, vertical stabilizers, rudder sections, spoilers, aerlerons, trailing edge panels, and access blowers in its DC-10 aircraft.

The payload door of the Space Shuttle orbiter, measuring 18.3 × 3.7 m, was the largest assembled structure when it was constructed. A boron-epoxy composite was used to fabricate a space truss for NASA. All Lunar missions used filament-wound and composite tanks for pressurized gases. In 1990, 18,000 tons of reinforced plastics were used for aircraft in the U.S.

## LAND TRANSPORTATION APPLICATIONS

In 1990, 312,000 tons of reinforced plastics were used for transportation applications in the U.S. The successor to the Jeep, called the Hummel (high-mobility multipurpose wheeled vehicle), has a hood, grille, assembly doors, and covers for the engine and battery made from fiber-reinforced plastic. The aerodynamics of motor homes and recreational vehicles has been improved by the use of FRP for front-end assembly, air dam, grille, and body panels.

Despite many years of success with the "hand lay-up" FRP body of General Motors' Corvette, little interest in this modern method of automobile construction was demonstrated until the oil crisis of the late 1970s. FRP truck cabs were made by "hand lay-up" techniques, and driveshafts and small automotive parts were made by pultrusion and SMC molding. However, the real breakthrough was the Pontiac Fiero, made by precision SMC molding of FRP.

The unprecedented Class A surface of the 250,000 Pontiac Fieros paved the way for the increased use of FRP bodies in the Chevrolet Camaro, Pontiac Firebird, and Ford Sierra, the entire floor pan of the Ford Escort, and transverse leaf springs, window frames, and hood in G.M.'s Cadillac. Wheels of the Fiero and racing cars, body panels and doors in the Tamworth Reliant Simitar SS, and personal rapid transit vehicles are made from FRP. The weight of plastics in an average American automobile increased from 77 kg in 1976 to more than 136 kg in 1989, and this volume will continue to increase. Production of the Fiero was discontinued because of motor problems, but the FRP body was recognized as a very successful composite. The Saturn currently utilizes reinforced plastics in its construction.

Filament-wound FRP is being used for tanker trucks carrying hazardous materials. The reinforced polyamide-imide Polimotor has performed satisfactorily in tests and on racing tracks. Thirty American universities with a total of 40 full-time faculty members are currently investigating FRP, but most are concerned with product design. More effort needs to be devoted to an understanding of the fundamentals of FRP. The growth of plastic composites is being hampered by a lack of knowledge of polymeric composite science.

## MARINE APPLICATIONS

Fiber-reinforced plastic was used for building boats in the early 1940s and is still the material of choice for marine applications. FRP boats used by the U.S. Navy vary in size from 3.7 m wherries to 17.4 m minesweeper hulls. The British ship *H.M.S. Wilton*, launched in 1972, was 46.7 m in length. Other large composite surface ships include the 24.4 m hydrofoil U.S.S.R. river passenger boat, the 23.4 m Netherlands pilot boat, and the 25.3 m African fishing trawler. The aramid-reinforced epoxy resin composite ship *The Flying Dutchman*, which received a gold medal at the 1984 Olympic Games at Los Angeles, CA, was named "Yacht of the Year" by *Yacht Racing and Cruising Magazine*. In 1990, 166,000 tons of reinforced plastics were used for marine applications in the U.S.

## FRP STRUCTURES

One of the first applications of fiber-reinforced plastic was as corrugated sheet in 1947. The use of this sheet increased 20-fold during the following five years and was supplemented by flat sheet in 1955. Monsanto's FRP "House of the Future" at Disneyland in the early 1950s demonstrated the advantages of FRP in construction.

The optimization of large FRP structures with constraints on strength and local stability, computer modeling of failure processes under load, and the strain developed in an aramid-reinforced plastic beam have been investigated. Apple and IBM have taken advantage of the electronic transparency of FRP and have constructed buildings with pultruded beams for testing laboratories.

A 20 m FRP sculpture weighing 3.2 tons has been erected on the waterfront at Boston, MA, and a 220 m tall power chimney has been erected at Bellingham, WA. New building concepts are now possible because of the utilization of filament-winding and pultrusion techniques. The states of Ohio and Florida have replaced metal highway signs with FRP signs. Chemical process equipment is being constructed from FRP.

New tests are also available for estimating the working life of FRP under load through the use of Zhurkov's equation. Deflection-strain relationships have been used to quantify the effect of flexing buried FRP pipe.

Plastics have usually been preferred for the construction of communications equipment. This use has been extended to filament-wound and pultruded FRP. Advanced composites of FRP are being used for television antenna satellite receivers. In 1990, 166,000 tons of reinforced plastics were used for corrosion-resistant applications in the U.S. The volume of reinforced plastics used in the U.S. in 1990 for electrical, consumer, and appliances were 106,000, 72,000, and 70,000 tons, respectively.

## RECREATION EQUIPMENT

The genesis of the modern vaulting pole is a story of advances made possible in the age of composites. The original pole was a natural composite of bamboo. Because of variances in quality, the bamboo was replaced by Swedish steel, which, because of excess rigidity, was replaced by aluminum. Vaulting records of above 5 m were broken frequently after the introduction of an FRP vaulting pole. Sergei Bobka of the U.S.S.R. has leaped over 6 m using this modern pole. While no records have been set, a comparable revolution has taken place with graphite-reinforced tennis rackets, golf shafts, and baseball bats.

Because of high cost and frequent breakage, many minor and amateur baseball leagues have replaced their wooden baseball bats with more permanent aluminum bats. However, because they do not have the characteristic sound of wood when hit by a baseball, these bats have not been accepted by the major leagues in the U.S. Aluminum bats are now being displaced by a blend of polycarbonate and polybutylene terephthalate (Xenoy) reinforced by graphite, which has the characteristic sound of a wooden bat.

# CONCLUSIONS

Advances in plastic composites technology which heretofore were considered breakthroughs are now routine. Boston Edison has constructed twelve 48-ft-diam wastewater treatment tanks which are 32 ft high, and more than a million leaking metal underground storage tanks are being replaced by reinforced plastic tanks. More than 7 tons of plastic composites are used in the Air Force C-17 transport, and advanced plastic composites that are stiffer, stronger, and lighter in weight than classic materials are used as helicopter blades for the V-22 Osprey tilt-rotor aircraft and the CH-47 cargo helicopter.

The Beechcraft Star Ship I is constructed from epoxy composites. Azdel thermoplastic sheets, which can be thermoformed, are being produced in a joint venture by PPG and G.E. FRP rods (Polystal), which are produced by the pultrusion of 83% glass/17% polyester resins, are being used to replace steel cable in stressed Portland cement structures. Additional developments will be noted as the use of compatibilizers increases.

Nostradamus (1503-1566) has fascinated scholars with his book of 906 predictions. However, neither he nor early 20th century prophets predicted the Age of Composites, which has made possible many things considered impossible a few decades ago. Few materials have affected modern civilization as much as polymer composites. This dynamic industry has improved aerospace, marine, and land transportation, construction, communications, and recreation and will continue to upgrade these and other material application areas in the future. In 1990, 1.14 million tons of reinforced plastics were used by the American plastics industry.

# REFERENCES

- B.D. Agarwal and L.J. Broutman, *Analysis of Performance of Fiber Composites*, Wiley-Interscience, New York, 1980
- P.F. Bruins, Ed., *Polyblends and Composites*, Interscience, New York, 1970
- R. Burns, *Polyester Molding Composites*, Marcel Dekker, New York, 1982
- L.A. Carlsson and R.B. Pikes, *Experimental Characterization of Advanced Composite Materials*, Prentice-Hall, Englewood Cliffs, NJ, 1987

- J. Garrison, (Epoxy resins), *Modern Plastics*, Vol. 66 (No. 11), 1989, p. 144
- M. Grayson, Ed., *Encyclopedia of Composite Materials and Components*, Wiley-Interscience, New York, 1983
- J.I. Kroschwitz, *Encyclopedia of Polymer Science and Engineering*, Vol. 5, Wiley-Interscience, New York, 1985
- G. Lubin, Ed. *Handbook of Composites*, Van Nostrand Reinhold, New York, 1982
- G. Lubin, Ed. *Handbook of Fiberglass and Advanced Plastics Composites*, Robert E. Kreiger Publishing, Huntington, NY, 1975
- P.K. Mallick, *Fiber Reinforced Composites*, Marcel Dekker, New York, 1988
- J.G. Mohr, S.S. Olesky, G.D. Skook, and L.S. Meyer, *Technology and Engineering of Reinforced Plastics/Composites*, Robert E. Krieger Publishing, Malabar, FL, 1984
- P. Morgan, *Glass Reinforced Plastics*, Cliffe and Sons, London, 1986
- T.J. Reinhart, Ed., *Engineered Materials Handbook*, Vol. 1, *Composites*, ASM International, Metals Park, OH, 1987
- R.P. Sheldon, *Composite Polymeric Materials*, Applied Science, London, 1982
- A.A. Watts, Ed. *Commercial Opportunities for Advanced Composites*, American Society for Testing and Materials, Philadelphia, 1980
- V. Wigotsky, (Aircraft composites), *Plast. Eng.*, Vol. 44 (No. 10), 1988, p. 25
- A.S. Wood, (Aircraft composites), *Modern Plastics*, Vol. 66 (No. 3), 1989, p. 40

# CHAPTER 12

# Tests for Plastic Composites

## INTRODUCTION

The pioneer designers, fabricators, and users of plastic composites depended to a large extent on manufacturers' claims and case histories in their selection of materials. However, there are now many tests established by standards organizations which ensure satisfactory performance of composites.

The largest standards organization is the International Standards Organization (ISO), which consists of members from 89 countries and many cooperative technical committees. In addition, there is the American National Standards Institute (ANSI) and the American Society for Testing and Materials (ASTM), which publishes its tests on composites annually in its Part 36, which includes reports from plastics committee D20. Other important reports on tests and standards are published by the National Electrical Manufacturing Association (NEMA), Deutsches Institut für Normung (DIN), and the British Standards Institute (BSI).

## EVALUATION OF TEST DATA

Unlike physical data available for metals, data for polymers are dependent on the life span of the test, the rate of loading, temperature, preparation of test specimens, etc. Some, but not all, of these factors have been taken into account in obtaining the data listed in tables in previous chapters of this book. Published data may vary for the same polymer fabricated with different equipment, produced by different firms, and for different formulations of the same polymer or composite. Hence, the values cited in the tables were labeled "Properties of Typical Polymers."

Many tests used by the polymer industry are adaptations of those developed previously for metals and ceramics. None is so precise that it can be used with 100% reliability. In most instances, the physical, ther-

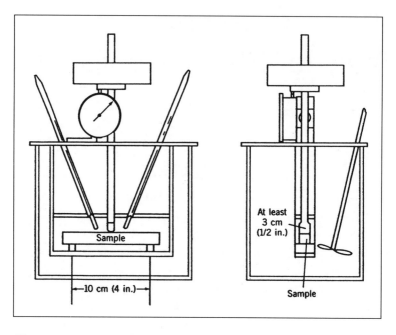

*Fig. 12.1 Heat deflection under load test*

mal, and chemical data are supplied by the producers, who are expected to promote their products in the marketplace. Hence, in the absence of other reliable information, such data should be considered as upper limits of average test data, and an allowance of at least ±5% should be made by the user or designer.

## HEAT-DEFLECTION TEST:
## ANSI/ASTM D648-72/78

This standard, which is now called "Deflection Temperature of Plastics under Flexural Load" (DTUL), is a result of "round robin" testing by all interested members of ASTM Committee D20. This heat-distortion standard was accepted several decades ago. As shown by the numbers after D648, it was revised and reapproved in 1972 and approved again in 1978.

This DTUL test measures the temperature at which an arbitrary deformation occurs when molded or sheet plastic specimens are subjected to an arbitrary set of testing conditions. The standard molded test span measures 127 mm in length, 13 mm in thickness, and 3 to 13 mm in width. The specimen is placed in an oil bath under a 0.460 or 1.820 MPa load in the apparatus shown in Fig. 12.1. The temperature is raised at a rate of 2 °C/min, and the temperature is recorded when the specimen deflects by 0.25 mm.

Since crystalline polymers, such as nylon 66, have a low heat-deflection temperature value when measured under a load of 1.820 MPa (264 psi), this test is often run at 0.460 MPa (66 psi). However, the 1.820 MPa load is standard for composites, and this value has been reported in the tables in this book.

The results of this test must be used with caution. The established deflection is extremely small and in some instances may be, at least in part, a measure of warpage or stress relief. The maximum resistance to continuous heat is an arbitrary value for useful temperatures, which are always below the DTUL value.

## COEFFICIENT OF LINEAR EXPANSION: ANSI/ASTM D696-79

Since it is not possible to exclude factors such as changes in moisture, plasticizer or solvent content, and release of stresses with phase changes, ASTM D696 provides only an approximation of the true thermal expansion (CTE). The values for thermal expansion of unfilled polymers are high, relative to those of other materials of construction, but these values are dramatically reduced by the incorporation of fillers and reinforcements. The CTE values will differ for longitudual and transverse tests when the composites are not isotropic.

In this test, the specimen, measuring between 50 and 125 mm in length, is placed at the bottom of an outer dilatometer tube and below the inner dilatometer tube. The outer tube is immersed in a bath and the temperature is recorded. The increase in length ($\Delta L$) of the specimen measured by the dilatometer is divided by the initial length ($L_o$) multiplied by the increase in temperature in order to obtain the coefficient of linear expansion ($\alpha$). The formula for calculating this value (CTE) is:

$$\alpha = \frac{\Delta L}{L_o \Delta T}$$

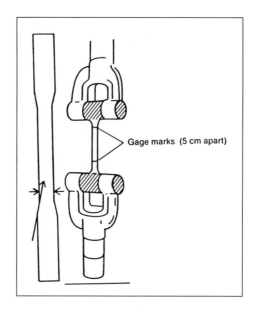

*Fig. 12.2   Diagram of a tensile test, showing the dumbbell-shaped specimen clamped in the jaws of an Instron tester*

## TENSILE STRENGTH:
## ANSI/ASTM D638-77 (D3039-76)

Tensile strength or tenacity is the stress at the breaking point of a dumbbell-shaped tensile test specimen 3.2 mm thick and with a cross section of 12.7 mm. The elongation or extension at the breaking point is the tensile strain. As shown in Fig. 12.2, the jaws holding the specimen are moved apart at a predetermined rate, and the maximum load and elongation at break are recorded. The tensile strength is the load at break divided by the original cross-sectional area. The percent elongation is the extension at break divided by the original gage length multiplied by 100. The tensile modulus is the tensile stress divided by the strain. As an alternative to reporting tensile strength/elongation, one may determine the slope of the tangent to the initial portion of the elongation curve. Typical stress-strain curves for polymer specimens are shown in Fig. 12.3.

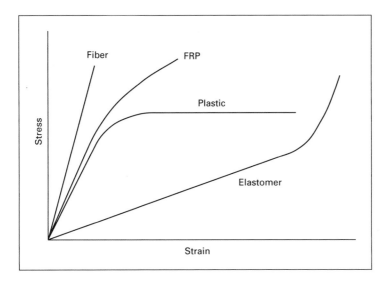

*Fig. 12.3 Typical stress-strain curve for fibers, fiber-reinforced plastics (FRP), plastics, and elastomers*

## FLEXURAL STRENGTH:
## ANSI/ASTM D790-81

Flexural strength, or crossbreaking strength, is the maximum stress developed when a bar-shaped test piece, acting as a simple beam, is subjected to a bending force perpendicular to the bar. An acceptable test specimen is one that is at least 3.2 mm in depth, 12.7 mm in width, and long enough to overhang the supports (however, the overhang should be less than 6.4 mm at each end).

The load should be applied at a specified cross-head rate, and the test should be terminated when the specimen bends or is deflected by 0.05 mm/min. The flexural strength (S) is calculated from the following expression, in which P is the load at a given point on the deflection curve, L is the support span, b is the width of the bar, and d is the depth of the beam:

$$S = \frac{PL}{bd^2}$$

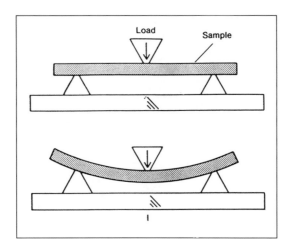

*Fig. 12.4 Effect of load on test bar in ASTM D790*

The following expression, in which D is the deflection, can be used to obtain the maximum strain (r) of the specimen under test:

$$r = \frac{6Dd}{L}$$

Data for the flexural modulus, which is a measure of stiffness, can be obtained by plotting flexural stress (S) versus flexural strain (r) during the test and measuring the slope of the curve obtained. The effect of the load (D) on the test bar in ASTM D790 is shown in Fig. 12.4.

## COMPRESSIVE STRENGTH:
### ANSI/ASTM D695-77

Compressive strength, also called compression strength, is the maximum stress that a rigid material will withstand under longitudinal compression. This strength is measured as force per unit area of the initial cross section of the test piece. The standard test specimen is a cylinder 12.7 mm in diameter and 25.4 mm high. The force of the compressive tool is increased by the downward thrust of the tool at a rate of 1.3 mm/min.

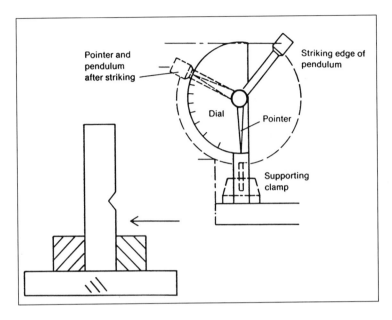

*Fig. 12.5 Notched Izod impact test (ASTM D256)*

## IMPACT TESTS:
## ANSI/ASTM D256-78

Impact strength may be defined as toughness or the ability of a rigid material to withstand a sharp blow, such as that from a hammer. The information obtained from the most common test (ASTM D256) on a notched specimen, as shown in Fig. 12.5, is actually a measure of the notch sensitivity of the specimen.

In the Izod test, a pendulum-type hammer, capable of delivering a blow of 2.7 to 21.7 J, strikes a notched specimen measuring 63.5 mm × 1.27 m × 1.27 m with a 0.025 m notch, which is held as a cantilever beam. The distance that the pendulum travels after breaking the specimen is inversely related to the energy required to break the test piece. The impact strength is calculated for a 25.4 mm test specimen.

In another test, called the Charpy impact test, the specimen is supported as simple beam. Values for unnotched specimens may be obtained in both the Izod and Charpy tests. Since plastics are usually brittle like glass, the impact values for unscratched specimens are always higher than those obtained for notched specimens. The Izod values are lower than the

Charpy values. There are also many less scientific, but more significant, tests such as those in which a ball is dropped from specified heights and those in which a plastic part is actually dropped onto a hard surface.

## HARDNESS TESTS

Hardness is dependent on the resistance of a material to local deformation. This test utilizes an indentor, which may be a sharp-pointed cone in the Shore D durometer test or a ball in the Rockwell test.

The Shore durometer is a spring-loaded indentor with a scale that shows the extent of indentation, with 100 being the hardest rating on the scale. The Rockwell tester measures the indentation of a loaded ball, usually on the R scale. The diameter of the ball used in the R scale is 12.7 mm.

## SPECIFIC GRAVITY (DENSITY)

Density is usually expressed as grams/cubic centimeter. Specific gravity is equal to the mass of a specific volume of the plastic compared to the mass of the same volume of water. Since the $g/cm^3$ units for the plastic material and water cancel out, specific gravity has no dimensions.

## REFERENCES

- B.D. Agarwal and L.J. Broutman, *Analysis and Performance of Fiber Composites*, John Wiley & Sons, New York, 1980
- *ASTM Annual Book of Standards*, Part 36, American Society for Testing and Materials, Philadelphia, 1986
- L.J. Broutman and R.M. Kroch, Ed., *Composite Materials*, Academic Press, New York, 1975
- J. Frados, *Plastics Engineering Handbook*, Van Nostrand Reinhold, New York, 1976
- R.M. Jones, *Mechanics of Composite Materials*, McGraw-Hill, New York, 1975
- J.I. Kroschwitz, Ed., *Encyclopedia of Polymer Science and Engineering*, Vol. 4, Wiley-Interscience, New York, 1986
- G. Lubin, Ed., *Handbook of Composites*, Van Nostrand Reinhold, New York, 1982
- P.K. Mallick, *Fiber-Reinforced Composites*, Marcel Dekker, New York, 1988

- T.J. Reinhart, Ed., *Engineered Materials Handbook*, Vol. 1, *Composites*, ASM International, Metals Park, OH, 1987
- R.B. Seymour, *Polymers for Engineering Applications*, American Society for Metals, Metals Park, OH, 1987
- R.B. Seymour and T. Cheng, Ed., *Advances in Polyolefins*, Plenum Publishing, New York, 1987
- R.B. Seymour and T. Cheng, Ed., *History of Polyolefins*, Reidal Publishing, Dordrecht, The Netherlands, 1986
- R.W. Tess and G.W. Poehlein, *Applied Polymer Science*, 2nd ed., ACS Symposium Series 285, American Chemical Society, Washington, DC, 1985
- W.V. Titow and B.J. Lanham, *Reinforced Thermoplastics*, John Wiley, New York, 1975

# U.S./SI Units:
# Definitions & Conversions

## STANDARD SYMBOLS

| | | |
|---|---|---|
| C = Celsius | | |
| = Centigrade | K = kelvin | Pa = pascal |
| g = gram | kg = kilogram | s = second |
| Hz = Hertz | m = meter | t = metric ton |
| J = joule | N = newton | |

| Numerical value | Term | Symbol |
|---|---|---|
| 10 | deka | da |
| $10^3$ | kilo | k |
| $10^6$ | mega | M |
| $10^9$ | giga | G |
| $10^{-2}$ | centi | c |
| $10^{-3}$ | milli | m |
| $10^{-6}$ | micro | μ |
| $10^{-9}$ | nano | n |

# METRIC CONVERSIONS

| To convert from metric or SI units | To U.S. units | Multiply by |
|---|---|---|
| **Length** | | |
| cm/cm/°C $\times 10^{-5}$ | in./in./°F $\times 10^{-5}$ | 0.555 |
| centimeter (cm) | inch (in.) | 0.394 |
| centimeter (cm) | foot (ft) | 0.0328 |
| meter (m) | foot (ft) | 3.2808 |
| millimeter (mm) | inch (in.) | 0.0394 |
| **Area** | | |
| $mm^2$ | $in.^2$ | 0.0016 |
| $m^2$ | $ft^2$ | 10.7639 |
| $cm^2$ | $in.^2$ | 0.155 |
| **Volume** | | |
| $m^3$ | $ft^3$ | 35.3147 |
| $m^3/kg$ | $in.^3/lb$ | 27,680 |
| **Mass** | | |
| gram (g) | pound (lb) | 0.0022 |
| kilogram (kg) | pound (lb) | 2.2046 |
| metric ton (t) | pound (lb) | 2204.6 |
| **Force** | | |
| MPa | psi | 145 |
| J/m | ft · lb/in. of notch | 0.0187 |
| N | lb/f | 0.225 |
| N/tex | g/denier | 11.33 |
| **Density** | | |
| $g/cm^3$ | $lb/ft^3$ | 62.43 |
| | | *(continued)* |

| To convert from metric or SI units | To U.S. units | Multiply by |
|---|---|---|
| **Temperature** | | |
| °C | °F | 1.8 °C + 32 |
| K | °F | 1.8 K − 459.67 |
| **Pressure** | | |
| kPa | psi | 0.145 |
| MPa | psi | 145 |
| GPa | psi | 145,038 |

## CONVERSION OF TABULATED DATA TO U.S. UNITS

| To convert from metric or SI units | To U.S. units | Multiply by |
|---|---|---|
| Heat-deflection temperature at 1.82 MPa, °C | at 264 psi, °F | 1.8 + 32 |
| Maximum resistance to continuous heat, °C | °F | 1.8 + 32 |
| Coefficient of linear expansion, cm/cm/°C × $10^{-5}$ | in./in./°F × $10^{-5}$ | 0.555 |
| Tensile strength, MPa | psi | 145 |
| Compressive strength, MPa | psi | 145 |
| Flexural strength, MPa | psi | 145 |
| % Elongation | % Elongation | 1 |
| Impact strength, cm · N/cm of notch (Izod) | ft · lb/in. of notch | 0.0187 |
| Impact strength, J/m | ft · lb/in. of notch | 0.0187 |
| Specific gravity | Specific gravity | 1 |
| Tenacity, N/tex | g/denier | 11.33 |
| Viscosity, Pa · s | poise | 10 |

# Glossary, Trade Names & Symbols

| | |
|---|---|
| A: | Symbol for a repeating unit in the polymer chain |
| ABA copolymers: | Block copolymers with three domains |
| ABFA: | Azobisformamide |
| ABS: | Acrylonitrile-butadiene-styrene |
| Accelerator: | Catalyst for crosslinking of rubber by sulfur |
| Acetal linkage: | Oxygen linkage |
| Acetal polymers: | $-[-CH_2O-]-_n$ |
| Acrylamide: | $H_2C=CHCONH_2$ |
| Acrylics: | Polymers based on acrylic acid esters; acrylonitrile or methacrylic acid |
| Acrylonitrile: | $H_2C=CHCN$ |
| Addition polymerization: | Chain polymerization resulting from addition of repeating units (mers) |
| Adipic acid: | $HOOC(CH_2)_4COOH$ |
| AIBN: | Azobisisobutyronitrile |
| Aldehyde: | $RHC=O$ |
| Alkanes: | $H(CH_2)_nH$ |
| Alkyd: | Reaction product of phthalic anhydride, a glycol, and usually an unsaturated oil |
| Alkyl group: | $H(CH_2)_n$ |
| Alternating copolymer: | A copolymer in which each repeating unit is joined to a different repeating unit in the polymer chain $(-A-B-A-B-)$ |
| AMBS: | Acrylonitrile-methyl methacrylate-butadiene-styrene |

| | |
|---|---|
| Amide group: | $-CONH_2$ |
| Amine (aliphatic): | $RNH_2$ |
| Anion: | Negatively charged atom |
| ANSI: | American National Standards Institute |
| Antioxidant: | Stabilizer against oxidation |
| Aramids: | Aromatic polyamides |
| Aspect ratio: | Length/diameter |
| ASTM: | American Society for Testing and Materials |
| Atactic: | A polymer in which the pendant groups have a random arrangement on both sides of the polymer chain |
| Atactic polypropylene: | Polypropylene with random arrangement of pendant groups |
| ATH: | Alumina trihydrate |
| B: | Boron; symbol for a repeating unit on the copolymer chain |
| Backbone: | The principal chain in a polymer |
| Balata: | *Trans*-polyisoprene |
| Bayer, O.: | Inventor of polyurethane |
| Bifunctional reactants: | Those with two functional groups |
| Bisphenol A: | $(HOC_6H_4)_2C(CH_3)_2$ |
| Block copolymer: | A copolymer consisting of long sequences of each repeating unit, like $A_nB_n$ |
| BMC: | Bulk molding compound |
| Bond angle: | The angle at which an atom is joined to another atom; this is $109.5°$ for C–C bonds |
| Bond length: | The average distance between two atoms; this is 0.154 nm for C–C bonds |
| Bonds, covalent: | Chemical bonds in which the electrons are shared |
| Bonds, single: | Bonds made up of two shared electrons |
| BPO: | Benzoyl peroxide |
| Branched polymer: | Polymer in which chain extension takes place at more than one position on the polymer chain |
| BSI: | British Standards Institute |
| Buna S: | SBR |
| Butyllithium: | $C_4H_9^-$, $Li^+$ |

| | |
|---|---|
| Butyl rubber: | Copolymer of isobutylene and isoprene |
| C: | Carbon atom |
| CAB: | Cellulose acetate butyrate |
| Carbanion: | Negatively charged organic ion |
| Carbonium ion: | Positively charged organic ion |
| Carboxylic acid: | RCOOH |
| Catalyst: | A substance which affects the rate of reaction but is not present in the product of the reaction |
| Catenated: | Atoms which are linked together, usually by covalent bonds |
| Cation: | A positively charged atom |
| CBA: | Chemical blowing agent |
| Celluloid: | Plasticized cellulose nitrate |
| Ceramics: | Materials based on baked clay |
| CFC: | Chlorofluorocarbon |
| Chain transfer: | A molecule from which an atom, such as hydrogen, may be readily abstracted by a free radical or other active species |
| Char: | Coke formation |
| Chardonnet, H.B.: | Coinventor of rayon |
| Chelate: | Five- or six-membered ring formation based on intramolecular attraction of H, O, or N atoms |
| Chemical blowing agent: | An agent that readily decomposes to produce a gas |
| Cl: | Chlorine |
| Cleavage: | Breakage of covalent bonds |
| Coefficient of expansion: | Increase in volµme/degree of temperature |
| Composites: | Polymers plus additives, usually reinforcements |
| Condensation polymerization: | The condensation of two difunctional reactants, usually with the elimination of a small molecule, such as $H_2O$ |
| Conformation: | Different shapes of polymers resulting from rotation about single covalent bonds in the polymer chain |

| | |
|---|---|
| Coordination catalysis: | Ziegler-type catalysis |
| Copolymer: | Macromolecule made up of more than one repeating unit |
| Coupling: | The joining of two macroradicals to produce a larger molecule |
| Coupling agents: | Difunctional compounds with attractive groups for fillers and resins |
| CPE: | Chlorinated polyethylene |
| cps: | Cycles per second |
| CR: | Neoprene |
| Creep: | Movement of a specimen under stress over a long period of time |
| CRIM: | Conventional reaction injection molding |
| Critical chain length: | Minimum chain length required for entanglement of polymer chains |
| Crosslinked polymer: | Three-dimensional polymer |
| Crystalline polymer: | Any polymer containing crystalline areas |
| Crystals: | Solids with characteristics based on the ordered arrangement of atoms or molecules in a definite pattern |
| CTE: | Coefficient of thermal expansion |
| Cumulative: | Additive or summation of effects |
| D: | Density |
| DAF: | Diallyl fumarate |
| DAM: | Diallyl maleate |
| DAP: | Diallyl phthalate |
| Degree of polymerization: | Number of repeating units in a polymer |
| Density: | Mass per unit value |
| Diadic: | Polyamide produced by the condensation of a diamine and a dicarboxylic acid |
| Diatomaceous earth: | Finely divided silica from skeletons of diatoms |
| Difunctional reactants: | Reactants with two functional groups |
| Diisocyanate: | $R(CNO)_2$ |
| Dilatometer: | Device for measuring changes in volume |
| Dimer: | Oligomer with DP of 2, i.e., $M_2$ |

| | |
|---|---|
| DIN: | Deutsches Institut für Normung |
| Disproportionation: | Termination by chain transfer between macroradicals to produce a saturated and an unsaturated polymer molecule |
| DMAU: | Dimethylaminourethane |
| DMF: | Dimethylformamide |
| DOP: | Dioctyl phthalate |
| DP: | Degree of polymerization; i.e., the number of repeating units in a polymer molecule |
| DPG: | Diphenylguanidine |
| Drawing: | Extending or stretching |
| Drier: | Organic salt of a heavy metal |
| DTUL: | Deflection temperature under load |
| E: | Energy |
| Ebonite: | Hard rubber |
| EEA: | Ethylene-ethyl acrylate copolymer |
| EGG equation: | $\eta = \eta_o(1 + 2.5C + 14.1C^2)$ (Einstein, Gold, Guth) |
| Elastomer: | A rubberlike polymer with high extensibility and low intermolecular force |
| Electromagnetic interference: | Interference related to accumulated electrostatic charge in a nonconductor |
| Electron: | A negatively charged lightweight particle |
| Electron, valence: | An electron in the outer shell of an atom |
| Ellis, Carlton: | Coinventor of unsaturated polyesters |
| EMF: | Electromagnetic force |
| EMI: | Electromagnetic interference |
| Emulsion polymerization: | Polymerization of monomers dispersed in an aqueous emulsion |
| Enthalpy: | $H$ = heat content |
| Entropy: | $S$ = a measure of disorder = 0 at 0 K |
| EP: | Epoxy resin |
| EPA: | Environmental Protection Agency |
| EPDM: | Crosslinkable ethylene-propylene copolymer |
| Epoxy resin: | Reaction product of bisphenol A and epichlorohydrin |
| EPS: | Expanded polystyrene |

ETDA: Ethyltoluene-diamine
Ethanol: $C_2H_5OH$
Ethylene: $H_2C=CH_2$
EVA: Ethylene-vinyl acetate copolymer
EVAL: Ethylene-vinyl alcohol copolymer
Fabrication: Conversion of polymer to finished article
Fibers: Threadlike, crystalline materials with extremely strong intermolecular attractions, which favor good fitting of adjacent polymer chains
Filament: Continuous fiber
Filament winding: The winding of a resin-impregnated filament on a mandrel
First-order transition: $T_m$ = melting point
Flexural strength: Bending strength
Folded chain: A crystalline polymer in which the polymer chain is folded back and forth
Free energy: $\Delta G = \Delta H - T\Delta S$
Free radical: An electron-deficient molecule or atom
Free rotation: The rotation of atoms, particularly carbon atoms, about a single bond; since the energy requirement is only a few kcal, the rotation is said to be free
FRP: Fiber-reinforced polymer
Full contour length: The length of a fully extended polymer chain
g/denier: 0.0088 N/tex
Gel coat: Unfilled outercoat in a laminate
GPa: 145,086 psi
GPC: Gel permeation chromatography
Graphite fibers: Carbon fibers
Guayule: Desert plant with high rubber content
Gutta percha: *Trans*-polyisoprene
$H^-$: Hydride ion
$H^+$: Proton
HALS: Hindered amine light stabilizer
HAR: High aspect ratio
Hardness: The ability of a substance to resist penetration or scratching
HBP: Hydroxybenzophenone

| | |
|---|---|
| HCFC: | Hydrofluorocarbon |
| HDPE: | High-density polyethylene, i.e., linear polyethylene |
| HDUL: | Heat-deflection temperature, under load |
| Heat-deflection temperature: | Temperature at which a loaded bar deflects a specified distance |
| Heat-distortion temperature: | Same as Heat-deflection temperature |
| *Hevea braziliensis*: | Natural rubber |
| Hexafluoropropylene: | $F_2C=C(CF_3)F$ |
| Hexa: | Hexamethylenetetramine |
| High-performance polymers: | Engineering polymers |
| HIPS: | High-impact polystyrene |
| Homologous series: | A series of organic compounds which differ by the number of methylene groups ($CH_2$) |
| Homopolymer: | A polymer name used to distinguish it from co-polymers |
| Hooke's Law: | $S = G\gamma$, stress = modulus × strain |
| Hydrocarbon: | A compound consisting of carbon and hydrogen atoms, like octane ($C_8H_{18}$) |
| Hydrogenation: | Addition of hydrogen to an unsaturated compound |
| Hydrogen bond: | The attraction between a hydrogen atom and an oxygen or nitrogen atom |
| Hydrogen cyanide: | HCN |
| Hydrolysis: | Decomposition by water |
| Hydrophilic: | Water-loving |
| Hydrophobic: | Water-hating |
| IIR: | Butyl rubber |
| Initiation: | The first step in a chain polymerization reaction |
| Initiator: | A substance that initiates a chain reaction |
| Inorganic: | Noncarbonaceous compound |
| Intermolecular attraction: | Attraction between atoms in different molecules |
| IR: | *Cis*-polyisoprene |
| ISO: | International Standards Organization |

| | |
|---|---|
| Isobutylene: | $H_2C=C(CH_3)_2$ |
| Isoprene: | $H_2C=C(CH_3)CH=CH_2$ |
| Isotactic propylene: | Propylene with pendant groups on one side of the chain |
| It: | Isotactic |
| Izod test: | Impact test |
| J/m: | 0.0187 ft · lb/in. |
| K: | Kelvin |
| Kaolin: | Clay |
| kcal: | Kilocalorie |
| K factor: | A measure of heat transfer |
| Kienle, R.H.: | Inventor of alkyds |
| kPa: | 0.145 psi |
| Lamellar: | Platelike in shape |
| LARC: | Thermoplastic polyimide |
| Latex: | Aqueous dispersion of a polymer |
| LCM: | Liquid composite molding |
| LCP: | Liquid crystalline polymer |
| LDPE: | Low-density polyethylene |
| LLDPE: | Linear low-density polyethylene |
| LOI: | Limiting oxygen index |
| London dispersion forces: | Weak intermolecular forces based on transient dipole-dipole interactions |
| LPA: | Low-profile agent |
| m: | Symbol for the molecular weight of a repeating unit |
| M: | Symbol for molecular weight |
| $M^+$: | Cation |
| $M^-$: | Carbanion |
| $M_c$: | Critical chain length |
| $M_n$: | Number average molecular weight |
| $M_w$: | Weight average molecular weight |
| Macrocarbonium ion: | Positively charged macromolecule |
| Macromolecule: | Extremely large molecule or polymer |
| Macroradical: | Polymerical radical $RM_nM\bullet$ |
| MBS: | Methyl methacrylate-butadiene-styrene |
| MBT: | 2-Mercaptobenzothiazole |

| | |
|---|---|
| MDI: | Methylene-4,4′-diphenyl diisocyanate |
| MEKP: | Methyl ethyl ketone peroxide |
| Melting point: | Temperature at which crystalline and liquid phases are in equilibrium |
| Methylene groups: | $-CH_2$ |
| Methyl rubber: | Poly 2,3-dimethylbutadiene |
| MF: | Melamine-formaldehyde resin |
| Microballoons: | Hollow spheres |
| Modulus: | Stress divided by strain, i.e., stiffness |
| Molding compound: | Mixture of polymer and additives ready for molding |
| Mole: | Weight of a molecule in grams |
| Monadic: | Polyamide produced from an amino acid or lactam |
| Monodisperse: | A polymer in which the molecular weights of all molecules are identical |
| Monofunctional reactants: | Those with a single functional group |
| Monomer: | Building block for polymers |
| MPa: | 145 psi |
| NBR: | Acrylonitrile-butadiene rubber |
| NEMA: | National Electrical Manufacturers Association |
| Network: | Crosslinked polymer |
| Novolac: | Condensate of phenol and formaldehyde under acid conditions |
| NR: | Natural rubber |
| N/tex: | 11.33 g/denier |
| Nucleation: | Crystal formation |
| Nylon: | A synthetic polyamide; repeating groups contain amide groups |
| Nylon 6: | Polycaprolactum |
| Nylon 66: | Polymer obtained by the condensation of hexamethylenediamine and adipic acid |
| O: | Oxygen |
| Opacity: | Property of being impervious to light |
| Organic: | Carbonaceous compound |
| Organic metals: | Conductive polymers |
| OSHA: | Occupational Safety and Health Administration |

| | |
|---|---|
| P: | Extent of reaction in Carothers' equation |
| PA: | Polyamide |
| PAA: | Polyacrylic acid |
| PAEK: | Polyarylether ketone |
| PAI: | Polyamide-imide |
| PAN: | Polyacrylonitrile |
| *Parathenium argentatum*: | Guayule |
| PB: | Polybutadiene; polybutene |
| PBA: | Physical blowing agent; polybutylene adipate |
| PBDPO: | Polybrominated diphenyl oxide |
| PBI: | Polybenzimidazole |
| PBT: | Polybutylene terephthalate |
| PC: | Polycarbonate |
| PCL: | Polycaprolactone |
| PCTFE: | Polychlorotrifluoroethylene |
| PDA: | *p*-Phenylenediamine |
| PE: | Polyethylene |
| PEEK: | Polyether ether ketone |
| PEI: | Polyether-imide |
| Pendant group: | A group attached to the main chain, such as the methyl group in polypropylene |
| Pentane: | $H(CH_2)_5H$ |
| Perfluorinated polymers: | Completely fluorinated polymers |
| Peroxy compounds: | Those containing O–O linkage |
| PES: | Polyether sulfone |
| PET: | Polyethylene terephthalate |
| Phase change: | Change from gas to liquid or liquid to solid, i.e., transition |
| Phenol: | $C_6H_5OH$ |
| Phillips' catalyst: | Catalyst based on $CrO_3$ supported on $SiO_2/Al_2O_3$ |
| Phosgene: | $COCl_2$ |
| Phosphazene: | Polymer with –N=P– backbone |
| Phosphorus pentachloride: | $PCl_5$ |

| | |
|---|---|
| Photoconductive polymers: | Polymers that conduct electricity in the presence of light |
| Photoinitiators: | Catalysts for UV-radiated polymerization |
| Physical blowing agent: | A gas, such as fluorocarbon or volatile hydrocarbon |
| PI: | Polyimide |
| PIS: | Polyimide sulfone |
| Plasticizers: | Flexibilizing additives |
| Plastics: | Molecules that can be molded to produce useful rigid solids; these molecules have moderately strong intermolecular forces |
| Plastisol: | A dispersion of polymers in a liquid plasticizer |
| PMF: | Processed mineral fiber |
| PMMA: | Polymethyl methacrylate |
| Poly 2,3-dimethyl-butadiene | Methyl rubber |
| Polyacrylamide: | Polymer with repeating unit $-[-CH_2-CHCONH_2-]_n$ |
| Polyacrylonitrile: | Polymer with repeating unit $-CH_2CHCN-$ |
| Polyamic acid: | Soluble precursor to insoluble crosslinked polymers |
| Polyarylate: | Aromatic polyester |
| Polydisperse: | A polymer in which the molecular weights of the polymer molecules are different |
| Polydispersivity index: | $M_w/M_n$ |
| Polyethylene: | Polymer of repeating ethylene units $-CH_2CH_2-$ joined together in a long chain |
| Polyfluorocarbon: | Fluorocarbon polymer, such as PTFE |
| Polymer: | An extremely large molecule consisting of a multitude of repeating units. The terms "giant molecule," "macromolecule," and "polymer" can be used interchangeably. The atoms in most polymers are joined by covalent bonds. |

| | |
|---|---|
| Polymer chain: | The backbone of a polymer, which usually consists of a series of covalently bonded carbon atoms |
| Polyphenylene sulfide: | Polymer with repeating unit $-[-C_6H_4-S-C_6H_4-]_n$ |
| Polyphosphazene: | Phosphonitrilic polymer |
| Polypropylene: | Polymer with repeating unit $-CH_2CH(CH_3)-$ |
| Polystyrene: | Polymer with repeating unit $-[-CH_2-CH(C_6H_5)-]$ |
| Polyvinyl alcohol: | Polymer with repeating unit $-CH_2CHOH-$ |
| Polyvinyl chloride: | Polymer with repeating unit $-CH_2CHCl-$ |
| POM: | Polyacetal, polyoxymethylene |
| PP: | Polypropylene |
| PPE: | Polyphenylene ether |
| PPES: | Polyphenyl ether sulfone |
| PPO: | Polyphenylene oxide |
| PPS: | Polyphenylene sulfide |
| Prepolymer: | Low molecular weight linear polymer |
| Primary covalent bonds: | Bonds between atoms |
| Processing: | Compounding, i.e., mixing of polymer with additives |
| Propagation: | Chain growth |
| Proton: | $H^+$ |
| PS: | Polystyrene |
| PSO: | Polysulfone |
| PSU: | Polysulfone |
| PTFE: | Polytetrafluoroethylene |
| Pultrusion: | The process in which a resin-impregnated bundle of filaments is pulled through a heated die |
| PUR: | Polyurethane |
| PVA: | Polyvinyl alcohol |
| PVAc: | Polyvinyl acetate |
| PVC: | Polyvinyl chloride |
| PVDC: | Polyvinylidene chloride |
| PVDF: | Polyvinylidene fluoride |
| PVF: | Polyvinyl fluoride |
| Pyrolysis: | Degradation at high temperatures |
| RCOOH: | Carboxylic acid |

| Reaction injection molding: | One in which polymerization occurs in the mold |
|---|---|
| Repeating unit: | The monomeric unit in the polymer chain |
| Resole: | Condensation of phenol and formaldehyde under alkaline conditions |
| Resole resin: | Linear phenolic resin produced by alkaline condensation of phenol and formaldehyde |
| Reversible elongation: | Recoverable elongation |
| RF: | Reinforcing factor |
| RFI: | Radio frequency interference |
| RIM: | Reaction injection molding |
| $RNH_2$: | Primary amine |
| RRIM: | Reinforced reaction injection molding |
| RTM: | Resin transfer molding |
| RTP: | Reinforced thermoplastic |
| Rubber: | A polymer of isoprene; term also used for synthetic elastomers |
| SAN: | Styrene-acrylonitrile copolymer |
| $Sb_2O_5$: | Antimony oxide |
| SBH: | Sodium borohydride |
| SBR: | Styrene-butadiene rubber |
| SBS: | Styrene-butadiene-styrene |
| Secondary valence bonds: | Van der Waals' forces |
| Segmental motion: | Wriggling of polymer chain |
| Shore hardness: | A measure of depth of needle penetration in polymers |
| SiC: | Silicon carbide |
| Silanes: | $H(SiR_2)_nH$ |
| Silicone: | Polysiloxane |
| Siloxane: | $[Si-O]_n$ |
| Simulated: | Simplified formulas in which hydrogen atoms are omitted |
| SIS: | Styrene-isoprene-styrene |
| SMA: | Styrene-maleic anhydride copolymer |
| SMC: | Sheet molding compound |
| Smith, Watson: | Inventor of glyptals |
| SMMA: | Styrene methyl methacrylate |

| | |
|---|---|
| Softening point: | An empirical value related to $T_g$ |
| Solubility: | The extent to which a solute will dissolve in a solvent |
| Spherulitic: | Extent of aggregates of crystals present in a polymer |
| Spinneret: | A series of small holes |
| Spinning: | Passing a polymer solution or melt through a small hole |
| SR: | Synthetic rubber |
| SRIM: | Structural reaction injection molding |
| SSAS: | Synthetic sodium aluminum silicate |
| Stereospecific: | Polymers with specific arrangements of pendant groups in space, such as isotactic polypropylene |
| Strain: | Elongation resulting from applied stress |
| Stress: | Force exerted on an object |
| Styrene: | $CH_2=CHC_6H_5$ |
| Suspension polymerization: | Polymerization of suspended beads of monomers in water |
| Synergism: | Cooperative effect of two additives |
| Syntactic foams: | Polymer filled with hollow beads |
| $T_g$: | Glass transition temperature |
| $T_m$: | Melting point, first-order transition |
| TAC: | Triallyl cyanurate |
| Tacticity: | The arrangement of pendant groups in space |
| TCNQ: | Tetracyanoquinodimethane |
| TDI: | Tolylene diisocyanate |
| Tensile strength: | Resistance to pulling stresses |
| Terephthalic acid: | *p*-Phthalic acid |
| Termination: | End of the polymerization process |
| Tertiary amine: | $NR_3$ |
| TFE: | Tetrafluoroethylene |
| Thermal conductivity: | Conduction of heat |
| Thermal properties: | Properties related to temperature |
| Thermoset: | Crosslinkable or crosslinked |
| TMC: | Thick molding compound |
| Ton, metric (t): | 2204.6 lb |
| TPE: | Thermoplastic elastomer |

| | |
|---|---|
| TPO: | Thermoplastic polyolefin |
| TPU: | Thermopolyurethane elastomer |
| TPX: | Polymethylpentene |
| TSSC: | Toluenesulfonyl semicarbazide |
| UF: | Urea-formaldehyde resin |
| UHMWPE: | Ultrahigh molecular weight polyethylene |
| Ultraviolet light stabilizers: | Stabilizers against degradation in sunlight |
| V: | Volume |
| Van der Waals' forces: | Attractive forces between molecules |
| VHS: | Very high structure |
| Vinyl alcohol: | A nonexistent monomer ($H_2CCHOH$) |
| Vinylidene fluoride: | $H_2C=CF_2$ |
| Viscoelasticity: | Having the properties of an elastic solid and a viscous liquid |
| VLDPE: | Very low density linear polyethylene |
| Vulcanization: | Crosslinking, usually with sulfur |
| Wood flour: | Attrition ground wood |
| XMC: | X-type molding compound |
| XU 218: | Polyimide |
| Yarn: | Spun fibers |
| Ziegler catalyst: | Coordination catalyst based on $TiCl_3$ and $ET_2AlCl$ |

## TRADE NAMES

| | | | |
|---|---|---|---|
| Abson: | ABS | Alton: | PPS/PTFE |
| Aclar: | Polyfluorocarbon | Ameripol: | Polybutadiene |
| Aclon: | Polyfluorocarbon | Amidel: | Polyamide |
| Acrilan: | Acrylic fiber | Amilon: | Nylon 6 |
| Adiprene: | Polyurethane elastomer | Ampcoflex: | Polyvinyl chloride |
| Alathon: | Polyolefin | Araldite: | Epoxy resin |
| Alfane: | Epoxy resin cement | Ardel: | Polyarylate resin |
| | | Arloy: | PC/SMA blend |
| Alkor: | Furan resin cement | Arnite: | Polyethylene terephthalate |

Astrel: Polysulfone
A-Tell: Polyester ether resin
Bakelite: Phenol formaldehyde resin
Bayblend: ABS/PC
Beetle: Melamine-formaldehyde resin
Blendex: ABS
Butaprene-N: NBR
Cab-o-sil: Finely divided silica
Cadon: SMA terpolymer
Calendar: Liquid crystal polymer
Capram: Nylon 6
Capron: Nylon 6
Celanar: Polyester
Celanex: Polybutylene terephthalate
Celcon: Polyoxymethylene
Chemigum: Styrene-butadiene rubber
Collodion: Solution of cellulose nitrate
CR-31: An allylic resin
Crylor: Acrylic fiber
Cyanacryl: Polyacrylic elastomer
Cycolac: ABS
Cycoloy: PVC-ABS blend
Delrin: Acetal homopolymer
Derakane: Vinyl ester resin
Dexion: Polyester fiber
Durel: Polyarylate

Durez: Phenol formaldehyde resin
Ekanol: Polyarylate
Enant: Nylon 7
Epikote: Epoxy resin
Epoxylite: Epoxy resin
Fiberfax: Aluminum silicate fiber
Fluorel: Polyfluorocarbon
Formica: Phenolic laminate
Fortrel: Polyester fiber
Gafite: Polybutylene terephthalate
Grilene: Polyester fiber
Herculon: Polyolefin
HET anhydride: Chlorinated cyclic anhydride
Hetron: Unsaturated polyester
Hifax: Polyolefin
Hostalen: Polyolefin
Hostalite: PVC/CPE blend
Hycar: Acrylonitrile elastomer
Hypalon: Sulfochlorinated polyethylene
Hytrel: Aromatic polyester
Instron: Instrument used for measuring tensile strength
Isoplast: Polyurethane
Kalrez: Polyfluoro polymer
Kamax: Acrylic polymer

| | | | |
|---|---|---|---|
| Kapton: | Polyimide | Natsyn: | Polyisoprene |
| Kel F: | Polyfluorocarbon | Neoprene: | Polychloroprene |
| Kerimid: | Polyimide | Noryl: | Polyphenylene |
| Kevlar: | Aramid | | oxide |
| Kodel: | Polyester fiber | Nycor: | Nylon/ionomer |
| Korez: | PF cement | | blend |
| Kralastic: | ABS | Nyder: | Nylon 6 |
| Kraton: | Styrene-buta- | Nyrin: | Nylon 6 |
| | diene block | Oppanol: | Polyisobutylene |
| | copolymer | Panlite: | Polycarbonate |
| K Resin: | Styrene copoly- | Paracryl: | PVC/NBR blend |
| | mer | Paral: | Polyether elasto- |
| Krynar: | Polyacrylic elas- | | mer |
| | tomer | Paxon: | Polyethylene |
| Kydene: | Polyvinyl chlo- | Pebax: | Nylon block co- |
| | ride | | polymer |
| Kydex: | PVC/acrylic | Pellothane: | Polyurethane |
| | blend | Perlon D: | Polyurethane |
| Kynar: | Polyvinylidiene | Perlon U: | Polyurethane |
| | fluoride | Perspex: | Polymethyl |
| Kynol: | Polystyrene | | methacrylate |
| Lexan: | Polycarbonate | Petlon: | Polyethylene |
| Lucite: | Acrylic | | terephthalate |
| Lustran: | ABS | Petra: | Polyethylene |
| Lycra: | Polyurethane | | terephthalate |
| | "snap-back" fiber | Petron: | Polyethylene |
| Makroblend: | PC/PET blend | | terephthalate |
| Makrolon: | Polycarbonate | Plaskon: | Urea-formalde- |
| Marvinol: | ABS | | hyde |
| Melinex: | Polyester | Plexiglas: | Polymethyl |
| Merlon: | Polycarbonate | | methacrylate |
| Micarta: | Phenolic lami- | Pliovic: | Polyvinyl chlo- |
| | nate | | ride |
| Mindel: | ABS/PSO blend | Pocan: | Polyethylene |
| Minlon: | Nylon | | terephthalate |
| Mondur: | Polyurethane | Polyman: | ABS/PVC blend |
| Moplen: | Polypropylene | Prevex: | Polyarylether |
| Mylar: | Polyester | Profax: | Polypropylene |

| | | | |
|---|---|---|---|
| Proloy: | ABS/PC blend | Surlyn: | Ionomer |
| Quacorr: | Furan | Tedlar: | Polyvinyl fluo- |
| Radel: | Polyaryl ether | | ride |
| Refrasil: | Silica fiber | Teflon: | Polytetrafluoro- |
| Renflex: | PP/EPPM blend | | ethylene |
| Rilsan: | Polyamide fiber | Terylene: | Polyethylene |
| Rinthan: | Polyurethane | | terephthalate |
| Ropet: | PET/PMMA | Texin: | Polyolefin |
| | blend | Therimid: | Polyimide |
| Rovel: | ABS | Thiokol: | Polysulfide elas- |
| Roylas: | Thermoplastic | | tomer |
| | elastomer | Torlon: | Polyamide-imide |
| Rucothane: | Thermoplastic | Trevira: | Polyethylene |
| | elastomer | | terephthalate |
| Rynite: | Polyethylene | Tynex: | Polyurethane |
| | terephthalate | Ucandel: | Polysulfone |
| Ryton: | Polyphenylene | Udel: | Polysulfone |
| | sulfide | Ultem: | Polyether-imide |
| Santocel: | Finely divided | Ultradur: | PBT/elastomer |
| | silica | Ultraform: | POM/elastomer |
| Santoprene: | Thermoplastic | Ultrathene: | Polyolefin |
| | elastomer | Urac: | Urea-formalde- |
| Saran: | Polyvinylidiene | | hyde |
| | chloride | Valox: | Polybutylene |
| Sclair: | Polyolefin | | terephthalate |
| Selar: | Nylon/HDPE | Veloy: | PET/PBT blend |
| | blend | Vespel: | Polyimide |
| Silastic: | Silicone | Vibrathane: | Polyurethane |
| Sil-Temp: | Silica fiber | Victrex: | Polyether sulfone |
| Skybond: | Polyimide | Victroy: | Polyphenyl sul- |
| Solprene: | Polystyrene | | fone |
| | block copolymer | Videne: | Polyethylene |
| Spandex: | Polyurethane | | terephthalate |
| | "snap-back" fiber | Vinylite: | Polyester |
| Stanyl: | Nylon 46 | Vistalon: | PPDM |
| Styrofoam: | Polystyrene foam | Vistanex: | PIB |
| Styron: | Polystyrene | Viton: | Polyfluorocarbon |

| Vitramid: | Nylon 6 | Vyrene: | Polyurethane fiber |
| Vulco-prene A: | Polyurethane elastomer | Vythene: | PVC/PV blend |
| Vulkollan: | Polyurethane | Xydar: | Liquid crystal polymer |
| Vydine: | Nylon | Zeolite: | Aluminum silicate |
| Vydyne: | Nylon | | |
| Vynite: | PVC/NBR blend | Zytel: | Nylon |

# Index

Polyvinyl alcohol
  as antiblocking agent, 12
  fibers, 84
Polyvinyl chloride
  aluminum filler for, 53
  blends of, 117
  chlorinated, 161, 163, 164
  effect of coupling agent on, 29
  fillers for, 57
  foams, 99–100
  heat stabilization of, 3, 33–34, 35
  impact modifiers for, 34–36
  plasticization of, 4
  processing aids for, 42
  production of, 161, 163
  properties of, 164
Polyvinyl fluoride, 149, 150
Polyvinylidene chloride, 161
Polyvinylidene fluoride, 149, 150
POM. *See* Polyoxymethylene
PPG Company, 201
PPO. *See* Polypropylene oxide
Prevex, 106
Printing inks, 6
Processing aids, 42–43
Production, U.S.
  of ABS, 110
  of acetals, 168
  of butyl rubber, 137, 139
  of carbon black, 58
  of fiberglass, 71
  of glass spheres, 61
  of high-density polyethylene, 153
  of linear low-density polyethylene, 154
  of metal powders, 62
  of neoprene, 137
  of nylons, 169

  of phenolic resins, 127
  of plastic composites, 189
  of polyesters, 174
  of polyimides, 180
  of polypropylene, 158
  of polypropylene oxide, 107
  of polystyrene, 161
  of polyvinyl chloride, 163
Production, worldwide
  of acrylics, 148
  of epoxy resins, 125
  of natural rubber, 142
  of plasticizers, 42
  of titanium dioxide, 65
Prolastic, 116
PSO. *See* Polysulfones
PS. *See* Polystyrene
PTFE. *See* Polytetrafluoroethylene
Pulse, polymer blend, 104
Pultrez Ltd., 76
Pultrusion, 195
  for fabricating reinforced plastics, 75–76
Pump, CPVC, 163
PVC. *See* Polyvinyl chloride
PVDC. *See* Polyvinylidene chloride
PVDF. *See* Polyvinylidene fluoride
PVF. *See* Polyvinyl fluoride

Quartz, 63, 64
Quebrachetol, in latex, 1

Racon, 94
Radel, 183
Radio frequency interference, 6, 53
Reaction injection molding, 76